A SHORT
HISTORY
OF
NUCLEAR
FOLLY

A SHORT HISTORY OF NUCLEAR FOLLY

RUDOLPH HERZOG

Translated by Jefferson Chase

MELVILLE HOUSE
BROOKLYN · LONDON

A SHORT HISTORY OF NUCLEAR FOLLY

Originally published in German as *Der
verstrahlte Westernheld und anderer Irrsinn
aus dem Atomzeitalter* by Rudolph Herzog
Published under the Imprint Galiani Berlin

Copyright © 2012, Verlag Kiepenheuer & Witsch
GmbH & Co. KG, Köln / Germany

Translation © 2013 Jefferson Chase

Melville House
145 Plymouth Street
Brooklyn, New York 11201

mhpbooks.com

ISBN: 978-1-61219-173-7

Printed in the United States of America
1 2 3 4 5 6 7 8 9 10

A catalog record for this title is available from the Library of Congress.

Contents

Preface

When I was twelve or thirteen years old, my uncle showed me the barbiturates he kept in his refrigerator. "I'll be swallowing them, when the big one comes," he told me. What he meant was the nuclear explosion that he was convinced would turn Germany into a radioactive desert. Ronald Reagan had been president of the United States since 1981, a cadre of unpredictable gerontocrats ruled the Soviet Union, and there was no end in sight to the lunatic arms race the two sides had been pursuing for the past forty years. Soviet jets had just shot down a Korean passenger plane over the island of Sakhalin, and the Americans were planning to install laser weapons in space. In Germany, responsible fathers were building fallout shelters in their backyards. Not even the movies offered a respite from the omnipresent worries about impending catastrophe. One of the big blockbusters of 1983, *The Day After*, revolved around the radiation-poisoned survivors of a nuclear holocaust, *When the Wind Blows* featured similar images in animated form, and *Atomic Café* mocked the hapless civilian measures ordinary people were supposed to take in the event of a nuclear war.

Everyone knew that Germany was directly in the crosshairs of the atomic conflict. The world's superpowers had agreed that Central Europe was to be the main battlefield, and our teachers made darn sure we were aware of that fact.

A German military officer who came to speak at my school told us about a NATO doctrine known as "flexible response." The flexible part was that the Allies could decide to respond to a potential attack by the Soviet Union with conventional or nuclear weapons. The officer couldn't explain how the tiny German *Bundeswehr* could ever hope to fight a conventional war against the Red Army, and he hinted that in case of "Communist" attack, a nuclear strike would be more or less unavoidable. The most likely course of history was leading directly into the abyss. American Pershing II missiles were already deployed in West Germany, ready to pay back people on the other side of the Iron Curtain for any acts of aggression.

Then, the Chernobyl disaster happened. The meltdown of the reactor destroyed any faith that atomic energy could be a boon to the world. The apocalyptic mood was complete. In Bavaria, where I lived then, some of the fallout from the accident came down to earth. We kids weren't allowed to drink milk, and the principal of my school ran around with a Geiger counter taking measurements. For days, when we went outside for recess, we were told not to sit on the ground—the principal thought that we would become contaminated. I can remember playing Frisbee with my friend Dirk in a farmer's field. A woman cycling past yelled at us for trampling on the crops, and Dirk responded that the potatoes were radioactive anyway.

Today, I can hardly imagine that scenes like these were once my reality. The generally depressed mood of the 1980s has dissipated, and the nuclear crosshairs are no longer focused on Germany, but on other parts of the world. The end of the Cold War brought with it new feelings of disinterested distance. People are beginning to forget the popular fears of nuclear war and total destruction from a bygone epoch. But the new sense of security among many in the West is misleading. The experiments of the Atomic Age continue to affect the world today, and many of the old problems have actually gotten worse. Among the legacies of previous decades are sites on every continent that have been permanently contaminated by atomic tests, uranium mining, or accidents with nuclear reactors and atomic waste. Meanwhile, the knowledge of how to build atomic weaponry continues to spread across the globe. Whereas before 1989 the primary danger was a nuclear war between the United States and the Soviet Union, nowadays the Middle East has become the main political powder keg. The majority of experts believe that Iran is developing nuclear weapons, and other countries, like Saudi Arabia, are developing civilian nuclear programs with ambiguous purposes. In the worst-case scenario, the Middle East will see a nuclear arms race, followed by cataclysmic war.

In light of what we know today about past and current dangers, it is astonishing how optimistic people once were about the civilian and military potential of atomic fission. This book depicts how, for a period of forty years, belief in human progress and planning for potential wars spurred on nuclear research, sometimes for the presumed benefit of humankind, sometimes for the destruction of enemies. To better illustrate this parallel process, I will focus on some of the more extreme examples of nuclear folly and foolishness, including hydrogen bombs that went missing and were never located, cowboys with radiation poisoning, and plutonium in space. My selection process is completely subjective, and I've chosen to ignore such familiar topics as Hiroshima, the Cuban Missile Crisis, Chernobyl, and Fukushima. My sources are recently released secret documents from the Cold War, personal reminiscences, official evaluations, and press reports from the period. They show how people in the past approached a new technology with a nearly fatal mixture of frivolity, naïveté, and unscrupulousness, and how they allowed economic and global-political interests to trump social and ecological reason. And as we examine past mishaps, it's impossible to avoid asking ourselves the unsettling question: Are people today really any more responsible than they were back then?

CHAPTER ONE

After the Bomb, the World's Most Dangerous Invention

Nuclear technology has been spreading around the globe for the past half-century. Some countries—like South Africa, Argentina, Libya, and Switzerland—started nuclear-weapons programs, only to cancel them. The International Nuclear Non-Proliferation Treaty recognizes only five "official" nuclear powers: the United States, Russia, China, France, and Great Britain. But there are four more "unofficial" members of the nuclear club who are not signees to that document. Israel developed nuclear weapons in the mid-1960s, and India tested its first atomic bomb in 1974. In 1998 and 2006, respectively, they were followed by two countries that don't have the best of reputations: Pakistan, which is considered a very weak state, and the unpredictable dictatorship that is North Korea.

Nuclear weapons have continued to spread, gradually but steadily, because of a small, unassuming machine, the centrifuge, which allows even atomic paupers to produce the material needed to build the bomb. In the Islamic Republic of Iran, thousands of centrifuges are currently in operation—ostensibly for the civilian purposes of producing atomic energy.

The story of this dangerous apparatus begins in the final days of World War II. Hitler's Germany was defeated, and the Soviet Red Army was marching upon the surreal remnants of bombed-out Berlin. The end of the Third Reich was simultaneously the beginning of a new political order. Parts of this transition happened publicly: the capitulation of Nazi Germany, the liberation of the concentration camps, and the division of the defeated enemy into occupation zones. But events that would set the stage for the Cold War were also transpiring away from the public eye. The United States and the Soviet Union had long seen themselves as rivals and were trying to gain advantages—cultural, scientific, and military— in a conflict between opposing systems. Both sides desperately wanted to profit from the know-how of German scientists and military technicians, and they specifically targeted experts with this aim in mind. Particularly coveted were members of the German "Uranium Association"—the researchers who had worked in Hitler's atomic program.

Disputes remain about how close German scientists got to building an atomic bomb for the Führer, but it is clear that in the early years of World War I, they had a significant lead over their American counterparts. Less than a year after Otto Hahn, Fritz Strassmann, and Lise Meitner discovered atomic fission in 1938, a Hamburg professor named Paul Harteck approached the Nazi War Ministry to discuss with the military leadership the possibility of constructing a nuclear weapon. The country that succeeded in building this doomsday device first, Harteck argued, would enjoy an enormous advantage over its enemies.

The Nazis had ample access to the materials needed to build the bomb. Germany contained huge uranium deposits, and heavy water—a crucial element needed for transforming uranium into weapons-grade plutonium—was being produced in occupied Norway. Yet although the crème de la crème of German physics, including Hahn, Carl Friedrich von

Weizsäcker, and Nobel laureate Werner Heisenberg, worked on the Nazi nuclear program, the Third Reich was unable to exploit its advantageous position.

It is impossible to reconstruct precisely why the uranium project failed in Nazi Germany. In all likelihood, a combination of factors, rather than a single reason, explain why Hitler never got the bomb. One popular explanation was that many of Germany's best scientific minds were Jews who had been expelled from the country and worked for the Allies during the war. Neither did the Nazi leadership do itself any favors by sending a number of researchers to the front, leaving the nuclear project understaffed. Furthermore, Heisenberg, who quickly became the head of the initiative, may or may not have intentionally sabotaged it. When asked by Armaments Minister Albert Speer how much money he needed to build an atomic bomb, Heisenberg asked for only 40,000 thousand Reichsmarks—while the Brookings Institute put the total cost of the U.S. atomic weapons program during World War II at 20 billion dollars. Heisenberg also dampened the enthusiasm of Speer and the military leadership by telling them in the summer of 1942 that the earliest a nuclear weapon would be ready was in three years. That prognosis made investing large amounts of energy in the project seem unattractive. Finally, targeted bombing raids and acts of sabotage by the Allies, for example on a shipment of heavy water that was being transported in Norway, spelled the end of productive nuclear research in Nazi Germany.

But probably the most important reason for the failure of the German nuclear-weapons program was the fact that the Nazi leadership, and above all Hitler, did not comprehend the destructive potential of an atomic bomb and thus did little to fast-track its development. Nuclear chemist Nikolaus Riehl, a "Uranium Association" insider, would later claim that German scientists had, consciously or not, refused to build the bomb for Hitler, even though they probably had the ability to do so. "A researcher or engineer," Riehl wrote in his memoirs, "who was driven by scientific curiosity and a love of technical experimentation would have hardly been able to withstand the lure of the uranium project, and the Germans would have gotten much further if they had been pressured to do so and supported in their efforts by the government." Riehl attributed the relatively "lukewarm" interest of Germany's rulers to the intellectual primitiveness of Hitler and his henchmen. Riehl writes: "They no doubt could understand things like missiles, which made a lot of noise and whose function was obvious. But they had no real comprehension of unfamiliar abstract concepts like the massive amount of energy that could be released by nuclear fission." In Riehl's view, insufficient state support convinced most researchers that "there would be no ultimate breakthrough in the uranium project before Hitler's downfall, so there was no need to consult their consciences."

The American scientists who were working on the Manhattan Project received contradictory signals from Germany. After a now-legendary meeting with Heisenberg in Copenhagen in September 1941, Danish physicist Niels Bohr brought a sketch of the German design for an atomic bomb with him to the United States. When his colleague Hans Bethe saw the sketch, he exclaimed: "My God, the Germans are planning to drop a nuclear reactor on London." J. Robert Oppenheimer, the father of the American bomb, is said to have merely remarked that as a weapon it was extremely "useless."

The controversial German historian Rainer Karlsch contends that German physicists, including Heisenberg's great rival Kurt Diebner, continued to work on a nuclear bomb even as the war was drawing to a close. Karlsch maintains that according to a report by Soviet spies, as late as March 1945, a few weeks before Germany's surrender, a mysterious explosion took place in the central German state of Thuringia. Trees were reportedly uprooted in a radius of 500 to 600 meters, several buildings were destroyed, and a number of POWs were killed. Karlsch concludes that this was the result of a nuclear test. But a search for residual radioactivity at the supposed test site in 2006 came back negative. The German governmental organization that examined the land in question therefore determined that no nuclear-weapons test had ever taken place there. And significantly, after the war, not a single prototype—to say nothing of a functional nuclear weapon—was ever found anywhere in occupied Germany. All that remained of the efforts of the "Uranium Association" was an experimental reactor discovered in a basement in the tiny town of Haiger-loch in the south of the country, and it was unlikely that this bit of machinery would ever have been capable of producing plutonium.

The only things of value left behind by the German nuclear program were the members of the "Uranium Association," and both sides of the Cold War were keen to exploit their know-how. But the efforts to recruit Nazi scientists were often farci-cally clumsy, as documents from a British special unit reveal. The so-called "T Force" was a mobile unit that moved out ahead of other Western troops and occupied German indus-trial concerns and research facilities. T Force soldiers abducted any research scientists they discovered and tried to get them to reveal what they knew with a combination of threats and incentives. But the T Force was unable to distinguish between valuable and worthless sorts of knowledge—on one occasion they interrogated a harmless German widow, extracting from her the formula for "4711," the original eau de cologne. Meanwhile, starting in 1943, the American intelligence services ran a program, code-named the "Alsos," that explicitly targeted German nuclear research. Whereas the original aim had been to determine the state of the German weapons program, by the end of the war it was charged with capturing Heisenberg and his colleagues and occupying atomic research facilities. On April 23, 1945, a vanguard expedition of the Alsos arrived at the research reactor in Haigerloch, which had been taken by French troops the day before. To keep the reactor from falling into French hands, the Americans immediately decided to dismantle and remove it. But the machinery contained neither uranium nor heavy water. In the end, the intelligence officers persuaded a group of German physicists to reveal where the missing substances had been hidden. The uranium was found buried in a field, the heavy water in the basement of an old mill.

The Alsos succeeded in capturing a number of major German scientists from April to October 1945, and the intelligence program was subsequently terminated. Thanks to their early interest, the Americans were better prepared for this task than were the other Allies. They were also aided by the fact that by the end of the war, most notable German researchers were located in the western half of the country. Scientists who had done atomic research at the Kaiser Wilhelm Institutes in Heidelberg and Berlin were detained in their apartments in Southern German towns, and Heisenberg himself was apprehended at his summer home on Lake Walchen in the Bavarian Alps.

The Soviets were at a great disadvantage when it came to recruiting German chemists and physicists, and to make matters worse for them, the United States dropped the first-ever atomic bomb on Hiroshima on August 6, 1945, demonstrating just how devastating a nuclear weapon could be. Stalin was deeply disturbed. The new weapon threatened to alter the equilibrium of power between the world's two aspiring superpowers. For a brief moment in history, the U.S. military appeared undefeatable, and the Soviet Union redoubled their efforts to build their own nuclear weapon, which had commenced in 1943. The head of Soviet atomic program, Igor Kurchatov, sent his finest researchers to East Germany to secure uranium and locate any remaining principals of the "Uranium Association." This operation, directed by State Security General Ivan Serov, was a resounding success, as the Soviets captured Nikolaus Riehl as well as physicist Robert Döpel, Director of the Berlin Kaiser Wilhelm Institute for Physical Chemistry Peter Adolf Thiessen, Nobel laureate Gustav Hertz, and electronics wizard Manfred von Ardenne.

Riehl later recalled that the scientists were treated very politely and that the "professionals" of the Russian State Security Service had been quite friendly:

> They gave me advice and sent chocolate, tobacco and other amenities my way. When I was transported to the Soviet Union, a particularly fearsome State Security lieutenant, a real brick wall of a fellow, ran after the car, shook my hand, wished me well and prophesied, "You'll soon be driving around Moscow in your very own car."

Once the prominent prisoners of war had arrived in the Soviet capital, they were quartered in the same luxurious villa where the German field marshal Friedrich Paulus and his staff had been kept after they surrendered at Stalingrad. On the dining hall walls, there was still a map upon which soldiers had traced the front lines. Riehl, Ardenne, and some of the others were even taken to the Bolshoi Theater, where Borodin's *Prince Igor* was being performed to celebrate the Soviet Union's victory over Nazi Germany.

The Soviets, of course, expected something in return for the hospitality shown to the captured scientists. That was made abundantly clear to them in a meeting with Lavrentiy Beria, Stalin's much-feared head of state security and the secret police. Ardenne summarized the tenor of that encounter, saying that Beria had told him, "You are now going to build bombs for us as well." Ardenne immediately recognized that he was trapped. If the Germans refused to cooperate with the Soviets, they would be sent to a labor camp or executed. If they cooperated, they would become privy to the most sensitive Soviet secrets and would never be allowed to return to Germany. Just moments after the challenge was posed, Ardenne hit upon a bold idea, which he then proposed to Beria and the other state-security operatives. "Building the bombs is the easy part," he told them. "You can do that yourself." But the German scientists would agree to produce the material needed to make a nuclear bomb.

In fact, the design of the bombs themselves was so comparatively straightforward that the Americans had not bothered to test a uranium bomb before dropping one on Hiroshima. "Little Boy," as the first nuclear weapon was nicknamed, contained two masses of uranium that were propelled into each other by a small "cannon shot" of explosives. The collision yielded a critical mass and unleashed a chain reaction of nuclear fission. Every time an atom was split, it emitted two neutrons, which would then split two further atoms, and on and on. In the process, enormous amounts of energy were released, with devastatingly destructive results.

A much more difficult matter than designing the bomb was extracting the fissible components of natural uranium. The key ingredient was the uranium 235 isotope, which made up a mere 0.72 percent of natural uranium. In Europe, at the time of Ardenne and Beria's meeting, researchers only had vague hypotheses about a process, called uranium isotope separation, that would allow them to produce tiny amounts of weapons-grade uranium. It had taken the United States years of massive industrial labor to collect the few kilograms of highly enriched uranium contained in Little Boy. The diffusion method the United States had used required tremendous amounts of energy and was extremely costly and inefficient. In fact, it was ill suited for a weapons program. Moreover, and even worse from the Soviet perspective, the details of this process were top secret.

Ardenne's promise to solve the problem of uranium isotope separation was a risky gambit. He suspected that the outcome would be very unpleasant if he and his team failed in their endeavor, which initially seemed to have more to do with alchemy than science. The Soviets debated Ardenne's proposal for half an hour before accepting it. Years later, at an official state reception, Soviet Premier Nikita Khrushchev took Ardenne to one side and congratulated him for how cleverly he had managed to wriggle out of the noose around his neck.

The Germans were then taken to Sinop, a picturesque little village around two kilometers from the Black Sea resort town of Sukhumi. There, a hotel complex had been hastily converted into a research institute and equipped with various bits of scientific hardware looted from German chemistry and physics laboratories. Anything further the researchers felt they needed was requisitioned from Soviet industrial facilities, although the scientists were no doubt warned of the dire consequences that would follow, should their experiments turn out to be a waste of time and resources.

Nonetheless, despite enormous efforts on all sides, the uranium project did not get off to an auspicious start. The key to success, Ardenne immediately realized, would not be technology, but a team of scientific minds capable of achieving the impossible. Frantically, Ardenne started searching POW camps for gifted chemists, physicists, and engineers, and Soviet authorities provided him with lists of specialists whom they had captured. Anyone who seemed interesting was whisked away to the sunny climes of the Georgian resort. Ardenne could not have suspected how lucky he would be. One of the two men who would prove crucial to the project just happened to be alive and well in a Soviet POW camp—and that by accident.

Max Steenbeck was the former director of the Siemens-Reiniger electrical and medical technology plant in Berlin. His groundbreaking research had earned him the nickname "the pope of plasma." As the Red Army advanced on the German capital, Steenbeck had defended the installation as a voluntary militiaman. Upon being captured by the Soviets, he swallowed a cyanide pill, but the poison had no effect. A friend who was also a chemist speculated that the digestive liquids in Steenbeck's stomach were abnormally basic and thus neutralized the effect of cyanide, which is an acid. The more probable explanation, though, is that the pill was a mere placebo.

After a forced march to the province of Posen, the unsuccessful suicide was interned in a POW camp. Steenbeck, a sensitive man who was used to enjoying the amenities of social status, had severe difficulty adjusting to the conditions of Soviet imprisonment. After a few weeks, his skin was covered with sores, and he was suffering from fever and persistent diarrhea. The only relief the seriously ill chemist was given was a straw sack as a blanket and a daily glass of milk. A short time after miraculously escaping death, Steenbeck was again very close to it. His condition was no doubt too poor for him to appreciate what was happening when Soviet officers appeared and took him away from his barracks in October 1945. In his memoirs, he recalled being "far too apathetic for any sensations," as he was led through the gate and the surrounding barbed wire, "a moment all of us had envisioned with hope and fear." Steenbeck still didn't comprehend the situation when he was taken to the camp commandant's office: "Nothing at all moved me, not even the fact that there was a table set before me with tasty things, no idea what they were—all I can remember is a glass of vodka. In my lethargy, I didn't think it was real. Everything seemed like a dream ... When I woke up the following day, I was lying in a real bed with fresh linens, all alone in a large room. I didn't really know why this should be the case." Steenbeck had been saved because his name was on Ardenne's list, although the "plasma pope" had no way of suspecting that at the time.

As soon as the Soviets had gotten Steenbeck back on his feet, he was taken to the Black Sea. The situation there was the opposite of the wretched life in the POW camp. The researchers under Ardenne may have been prisoners who had to obey instructions from Soviet authorities, but the Germans in Sukhumi scarcely noticed that their liberty was curtailed. The living conditions were nearly ideal, much better than those enjoyed by average Soviet citizens or people back at home in war-ravaged occupied Germany.

Sukhumi, today the capital of Abkhazia, is located at the foot of the Caucasus Mountains, and its broad beachfront promenade, which, sadly, was destroyed in the Georgian civil war of the 1990s, attested to the city's glamorous past. It was Beria's birthplace and, in Soviet days, a flourishing port known as "the white city on the sea." Sukhumi was part fishing village and part Soviet pomp, a pastoral spa town with Communist *grandezza*, a mixture that had an undeniable sultry charm. Stalin's security service resided under palm trees that were hundreds of years old. Next door, people could bathe in sulfur waters and stroll along the promenade. The authorities had built gigantic apartment buildings for the working-class elites, and on every floor of these Orwellian structures there was guaranteed to be a government spy. On weekends, deserving Soviet citizens would pour in and get rousingly drunk on the beach or take a romantic hike in the Caucasus that also usually ended in an alcoholic bender. Life in Sukhumi was as decadent as possible within the constraints of Stalinist dictatorship.

This was the surreal environment in which Steenbeck suddenly found himself. He'd been informed before his arrival that he was to help build a Soviet atomic bomb, and faced with no alternative, he'd agreed. Ardenne immediately made him a departmental director, and he was given a chauffeur-driven car and an expansive room with a park-front view. From the foyer of the villa, an open-air staircase flanked by lions carved in sandstone led out into the greenery. Photos from the 1940s and '50s give a good impression of how the Germans lived in Sukhumi: they show studies with bookshelves up to the ceiling, polished hardwood floors, and friendly-looking men smoking pipes on the veranda. The bosses resided in gleaming white villas, the engineers and laborers in purpose-built, rustic wooden houses. A brook ran through the middle of the research center, contributing to the "Riviera in the Caucasus" atmosphere described by one of the researchers:

Wild vines with sweet grapes cover the trees amidst the subtropical vegetation and along the brook over the grounds. Plums and other edible fruits grew there too, for anyone who wished to pick them. Mandarin oranges were cultivated on the surrounding estates, and you could buy 20-kilogram crates of them for a cheap price.

The only disturbances in this paradise were "the unbelievably loud and sleep-inhibiting concert of frogs and fish" in the brook.

In March 1946, a late arrival and the second key figure in the Soviet uranium project joined the Black Sea group. The Austrian engineer Gernot Zippe had the sort of practical mindset that perfectly complemented Steenbeck's extraordinary conceptual abilities. Zippe had amply demonstrated his technical acumen during World War II. He had helped develop the first German radar tracking system, and later he had experimented with new, ultra-fast airplane propellers at the Luftwaffe's research center in Prague. A fellow POW suggested that the Soviets take a closer look at this multifaceted technician.

In contrast to Steenbeck, who eventually became a committed socialist and later served as a member of the State Research Council in Communist East Germany, Zippe was no great admirer of Marx and Engels. In fact he freely admitted his abiding admiration for the Führer. In his memoirs he wrote:

On account of my education and my experience of the Engelbert Dollfuss and Kurt Schussnigg dictatorships in Austria, and notwithstanding our defeat in the war, I avowed that I was a passionate follower of Adolf Hitler ... Of course, there were things I, and not only I, found out "afterward." For example, we were shown films of the horrors of the concentration camps, of which I previously had no knowledge. But the fact was that I had been able to experience the unbelievable mood of optimism accompanied by a series of great social triumphs.

Zippe and Steenbeck not only differed radically in their political views; they were also two very different sorts of men. Whereas the Austrian was reserved and introverted, Steenbeck was considered an arrogant know-it-all. Even in his own book *Crisis and Renewal,* he described himself as egocentric. Yet despite these differences in outlook and temperament, the two men quickly formed the most efficient team within the Sukhumi uranium project. The key to their success was mutual respect. Steenbeck, who was nominally Zippe's boss, appreciated the latter's pragmatic intelligence and left him to his own devices. Zippe, for his part, acknowledged Steenbeck's visionary energy and spirit. That is not to say that their collaboration was without some rough patches. Significantly, Steenbeck hardly mentions the Austrian in his autobiography. Zippe's memoirs, by contrast, contain a number of critical asides aimed at Steenbeck. The brilliant theoretician, Zippe wrote, "often had his head so far in the clouds that you would think he believed that nature necessarily had to follow his calculations, whereas the opposite, in my opinion, is the case." Zippe claimed that he often had to bring Steenbeck "back down from his lofty dreams to the terra firma of natural laws and constraints."

It was the theoretician Steenbeck who hit upon a new

method of producing uranium 235. The idea was to construct a special centrifuge to isolate minute quantities of weapons-grade material from natural uranium. Steenbeck was hoping to exploit the fact that in its gaseous form, uranium 235 is a tiny bit lighter than the rest of the radioactive metal. When uranium hexafluoride was run through a centrifuge, uranium 235 would be concentrated in the middle. The principle was the same as with an everyday washing machine: clean water weighs less than dirty water and thus remains in the middle of the drum. In a washing machine the drum needs to revolve fifteen times a second to produce this effect. Steenbeck calculated that in order to be capable of separating uranium 235 from other isotopes, the drum of the centrifuge would have to revolve a dizzying, supersonic 90,000 times a minute. That presented an almost insoluble dilemma since speeds of that sort would rip apart any normal drum, no matter how solidly it was constructed. No one in 1946 knew how to build the type of centrifuge the project required. Nonetheless, Steenbeck's plan seemed to be the only viable option. A second research group in Sukhumi, which had been experimenting with diffusion methods, had made little progress, and a third procedure that Steenbeck had initially favored had proven utterly impracticable.

Meanwhile, the researchers' Soviet masters were increasing the pressure to succeed. Many of the scientists would later recall that Beria kept a list of names in his safe, detailing which researchers would be sent to labor camps and which executed in the event that the bomb project failed.

So Steenbeck and Zippe got down to work. Wherever they looked, all they saw were seemingly insurmountable obstacles. The first issue was one of material. As soon as the centrifuge drums reached a critical velocity, they bent out of shape and eventually disintegrated. No metal could withstand that sort of strain. Zippe and Steenbeck began experimenting with rubber tubing, but it developed what they called "vibration bellies" and burst. Experiments were also carried out with glass and brass—with perilous results. The cylinders invariably broke into countless splinters that would go flying through the laboratory. Zippe considered asking for a suit of knight's armor to wear for protection. "It was only down to a huge amount of luck," he later wrote, "that there was no serious accident during the first phase of experimentation." The Germans working on other parts of the project could only shake their heads every time there was an explosion in the centrifuge laboratory. Steenbeck later acknowledged that his group was considered "crazy." Their experiments were made even more dangerous because Zippe and Steenbeck filled their prototypes with uranium hexafluoride, which occurs as sparkling crystals in an airless void but is transformed at the slightest contact with water into a potent acid. Any contamination with water meant that the liquid would eat through the centrifuge drum, even if it was made of the strongest glass. Moreover, the unpredictability factor was dramatically increased by the fact that Zippe's Russian laboratory assistant habitually drank a full measuring cup of 192-proof alcohol before starting every shift. Safety precautions at Sukhumi ranged from lax to nonexistent. On one occasion, an explosion in another laboratory blinded one of Zippe and Steenbeck's colleagues. He was taken away by his assistants and never seen again.

Nonetheless, despite all the obstacles, Zippe and Steenbeck were making progress. They came up with the idea of using coupling links to prevent metal drums from breaking apart due to eccentric motion at critical velocities. Meanwhile, Steenbeck constructed an electric motor capable of rotating the drum at the necessary speed. Initial results were encouraging, yet there was still a decisive hurdle to be overcome: friction. Whenever two bodies rub against each other, heat is generated. In the case of a centrifuge spinning at supersonic speeds, the result was an extraordinary amount of heat that destroyed the machine. Even when the mechanical parts barely touched one another, friction with the surrounding air was enough to cause irreparable damage.

To relax and clear his head, Zippe went on extensive tours through the mountains. The captive scientists enjoyed four weeks of holiday, the division heads, six. Zippe would drive a Jeep through the Caucasian highlands, accompanied by a guard who also served as an interpreter. On a hike around Lake Riza, at an elevation of 3,200 meters, Zippe looked down upon fields of snow and plateaus upon which horses were galloping. Time was gliding by. At a time when the first regular POWs were being allowed to return to Germany, the scientists labored on in their golden cage on the Black Sea. Days turned into months, and months into years.

Then, on August 29, 1949, things changed abruptly. At 7:00 a.m. that day, the Soviet Union successfully tested its first atomic bomb—without the help of the Sukhumi centrifuge-makers. Kurchatov had simply bypassed the problem of producing uranium 235 by using another fissionable substance, plutonium. Plutonium naturally occurs only in the minutest quantities, but it can be created in nuclear reactors from the energy released when uranium 238 is bombarded with neutrons. The fuel needed to run the reactors is a lower grade of uranium 235, enriched only by a factor of 4 to 5 percent, compared with the 80 to 90 percent needed for an atomic bomb.

Thanks to a few well-placed spies within the American atomic-weapons program, Kurchatov had succeeded in overcoming the two major barriers to building a plutonium bomb. First of all, producing plutonium was a painstaking process, and the Soviets were forced to master the techniques of running huge reactors. Second, it was far more difficult to detonate a plutonium bomb than a uranium one. The forces working upon the radioactive material had to be absolutely regular in order to achieve an explosive chain reaction. The solution was to surround a plutonium ball with highly volatile conventional explosives, which were then detonated with an extremely complicated mechanism. That caused the plutonium to implode, triggering a critical mass like the one that resulted in the two halves of uranium 235 in the cannon-detonation procedure. The result, which Kurchatov witnessed with his own eyes, was a gigantic, multicolored fireball that rose into the sky over Semipalatinsk, Kazakhstan.

The test represented a historical watershed for the world as a whole and for the United States in particular. Mutual mistrust between the superpowers intensified, and it was clear that the Cold War had the potential of turning into a direct, global military battle inflamed by the destructive power of atomic fission. The West focused even more on what was happening at Sukhumi and the other Soviet nuclear research facilities. The Soviets had mastered the laborious techniques needed to make a plutonium bomb, but would they be able to build a uranium one? When Steenbeck's wife was granted permission to join her husband in Sukhumi, she was contacted by a member of the British secret service, MI6, who asked her to smuggle sensitive information back to the West inside her lipstick case. She politely refused. But the incident, together with an extensive CIA report, shows that Western intelligence agencies knew at least the outlines of what was going on in Steenbeck and Zippe's laboratory.

In the end, the centrifuge-makers achieved several break-throughs. To avoid the problem of excessive heat from friction, Zippe mounted his centrifuge much like a top on a tiny needle. The drum was held in place vertically with strong magnetic rings, which meant it did not need to be mounted in conventional fashion. The furious revolving cylinder did not even make contact with air, since the interior of the machine was a vacuum. In this way, a centrifuge could attain the necessary revolutions without being subjected to destructive friction.

The Soviet repaid the researchers for their help by giving them various state decorations and then released them from captivity in the early summer of 1956. Zippe and Steenbeck's records were confiscated, and even a notebook of sheet music was taken away from them because it seemed suspicious. On July 28, Zippe travelled back home to Austria via Moscow and Budapest. He didn't have a single note in his luggage, but his head was full of valuable information. As soon as they could, Zippe and Steenbeck took out a joint patent on their uranium centrifuge. Zippe subsequently arranged with the CIA to be brought to the United States, where he handed over detailed plans for the device he had envisioned in Sukhumi to dumb-founded American officials.

The advantages of the procedure Zippe proposed for enriching uranium were obvious. Centrifuges were reliable and energy-efficient. One no longer needed reactors to produce weapons-grade material. By using a sufficient number of centrifuges, employed one after another in so-called "cascades," it was possible to enrich a large quantity of uranium 235 in a short time. It was far less complicated to design a uranium bomb than a plutonium one, and the costs of a nuclear-weapons program oriented in this way were seductively low.

It is no accident that atomic weapons produced in part by Zippe's centrifuges became known as the "poor man's bombs." Last but not least, a bomb-building program that employed centrifuges was relatively easy to conceal from prying eyes—a fact that would become ever more significant with the passing of decades.

Most of the advantages of the centrifuge-built bomb also applied to the enrichment of uranium for civilian purposes, such as producing fuel for nuclear power plants. Zippe exploited this fact and went to work for the nuclear energy industry, joining the Degussa company in Frankfurt in the 1960s and later moving to the German-Dutch-British consortium Urenco. As part of the contracts he signed, the companies had to pledge that they would only apply his and Steenbeck's research to civilian purposes and would take every precaution to ensure it was not misused. Nonetheless, Zippe would have been hard pressed to answer, had he been asked, how a multinational corporation like Urenco, with hundreds of employees scattered in several locations around the world, could have possibly guaranteed it would safeguard the secrets of his centrifuges.

The need for uranium enrichment grew exponentially with the burgeoning of nuclear power in Europe, and that put Urenco under enormous pressure to deliver sufficient amounts of reactor fuel. To meet the rapidly rising demand, the company passed a number of responsibilities on to subsidiaries and partner companies, including the Dutch consultancy firm Fysisch Dynamisch Onderzoekslaboratorium (FDO). In 1973, Urenco managers complained about metallurgic problems in their uranium centrifuges, and FDO sent a young man named Abdul Qadeer Khan to deal with the problem. He was a Pakistani engineer. The professor who supervised his dissertation later described him as "a nice guy," and he was married to an equally intelligent and attractive South African woman. Khan was a moderate Muslim who made no secret about his patriotism toward his home country. The life he led in the Netherlands was unremarkable. He and his family lived in a terrace house in the Amsterdam suburb of Zwanenburg, and like many other people, the Khans enjoyed going to the seaside or making trips to the Ardennes on the weekends. No one noticed that Khan had lied when he applied for his job and falsely claimed that his wife Hedrina "Hennie" Khan was a Dutch citizen. During his time at Urenco, Khan had free access to the large centrifuge halls that the company maintained in the town of Almelo. A colleague named Fritz Werman later recalled that Khan often took blueprints back home with him, in violation of the strict confidentiality rules of his employer. In addition, Hennie, who like her husband was multilingual, translated sensitive Urenco documents from Dutch to English.

The rest of the story is widely known. A.Q. Khan was a Pakistani spy. When he returned to his homeland, he provided Pakistani officials with blueprints of uranium centrifuges and lists of companies that supplied Urenco with parts. A number of European companies, including FDO, had no qualms about earning huge profits assisting Pakistan's atomic program. FDO supplied nuclear know-how and centrifuge components, even though the employees of the company surely must have known what their former colleague Khan intended to do with them. Germany was the biggest supplier of trade-restricted material. The CES Kalthof company in Freiburg even sent Khan an entire enrichment facility, declaring it to customs authorities as a "toothpaste factory."

Pakistan carried out its maiden nuclear test on May 28, 1998. But that was just the first in a series of alarming developments. In the meantime, Khan was doing booming business with Zippe's centrifuges, whose secrets he passed on to North Korea, Libya, and Iran, among others. Everywhere around the world, dictators hoped they could acquire this useful and potentially deadly technology. At the same time, further sensitive information was leaking out of the Urenco consortium. Even Saddam Hussein was able to exploit German connections to procure uranium centrifuges. The brand name was still clearly visible when they were later discovered after the First Gulf War. And Khan's representatives negotiated for months with al-Qaeda after the terrorist group decided to pursue its own nuclear program.

The absolute low point, thus far, in this story came shortly after September 11, 2001, when then–head of the CIA George Tenet met with George W. Bush to inform the president that al-Qaeda had smuggled a functional atomic weapon into New York City and were planning to detonate it. It turned out that the CIA based its warning on false information from an agent codenamed "Dragonfly," but the incident provided a taste of the sorts of things the world can expert if centrifuge technology is allowed to proliferate. For example, the SILEX process (the Separation of Isotopes by Laser Excitation) has recently been advanced by a number of private companies and raises the possibility of a new era of proliferation. First developed in the 1990s by private Australian researchers, the process produces enriched uranium using lasers and is far more effective than traditional centrifuge enrichment and potentially even easier to conceal. In 2010, the Iranian regime claimed to possess technical knowledge of the process. German nuclear weapons expert Hans Rühle warned in 2012 that "laser uranium enrichment is so attractive that it will be implemented—and Iran could become the test case."

Gernot Zippe always pointed out that his invention could be used to both the benefit and the detriment of humanity. It was like a kitchen knife, he told the BBC four years before his death in 2008. The analogy was simple: "With a kitchen knife you can peel a potato or kill your neighbor."

CHAPTER TWO

The Red Bomb

In Russia in the early 1990s, you could take an old ferry with wooden decks and spacious, comfortable berths from St. Petersburg all the way to the northernmost regions of Europe. The ship was a throwback. Through the large, cream-colored portholes, you could watch first the city, then expansive fields, and finally the banks of Lake Ladoga drift by in the sunset. The light would fade and with it the view, and passengers would fall into a deep, dreamless sleep. The next day you would rise in a foggy, opaque, mysterious world. Even after the fog had lifted, there was no land in sight—only a single, milky-white, completely still surface. Three hours later, you would reach Valaam, a forested island that is the site of a famous fourteenth-century monastery. With the collapse of Communism, eight monks moved there and set about meticulously restoring the decrepit building. On holidays, pilgrims would make the journey by ferry from St. Petersburg and be given simple meals made in the monastery kitchen. If you wanted to, you could also fish in the teeming waters of Lake Ladoga.

Valaam was a magical place, like a mythical land on the other side of the rainbow. At the same time, the surrounding countryside, especially to the north, was one of the most heavily radioactive regions in all of Russia. This discovery was particularly alarming because Lake Ladoga supplies some of the drinking water for St. Petersburg's millions of inhabitants.

The contamination had taken place in the early 1950s, when a series of nuclear tests was carried out on Lake Ladoga's north shore. Two of the islands in the Valaam archipelago that were used for the tests were given new names. Heisägenmaa, formerly part of Finland, became Suri, while Makarinari was rechristened Maly. The new names were an attempt to confuse Western intelligence services. There was a variety of reasons why Soviet authorities decided to use a fresh-water reservoir as a nuclear test site. The upper northern reaches of Russia were largely uninhabited, and thus the tests were "safe," in the rather blinkered view of the Soviet military. The desire to extract a bit of revenge on Finland likely also played a role. Russian commanders remembered all too well the humiliating resistance that Russia's much smaller neighbor had put up against the Red Army in the Winter War of 1939–40. Soviet troops also tended to test out explosive devices in former Finnish military bunkers, although that choice may have been made for technical reasons and not out of revenge.

Most of these experiments involved dirty bombs, conventional explosive devices packed with radioactive material that did not set off a nuclear chain reaction. The objective was to determine how radioactivity could be dispersed and what effects it would have on living organisms. The Communist leadership, and in particular Stalin himself, feared that the United States would attack the Soviet Union, and at that point in the nuclear arms race, America still had the advantage in terms of the size of its arsenal. With a constant stream of hard-line anti-Communist rhetoric coming from the other side of the Atlantic, people in the USSR assumed the worst was nigh. Such assumptions were by no means completely wrong. In

1949, the U.S. military developed a plan code-named "Operation Dropshot" that foresaw destroying a hundred Soviet cities with nuclear weapons in the event that the Soviets launched a conventional ground offensive against Western Europe. The Soviets thus needed to know what consequences an atomic attack would have. Would the Red Army even be able to defend itself in a radioactively contaminated environment? In 1953, in order to answer such questions as well as to investigate the potential of radioactive weapons, the Red Army began to detonate radioactive bombs and other explosive devices in and on Europe's largest inland lake. No one was concerned that, as a result, radioactive water would be transported via a network of rivers as far away as the Gulf of Finland.

The test carried out aboard the *T12*, a German mine-layer, proved to be particularly problematic. In 1945, the vessel, which had been part of Hitler's navy, was surrendered to the Soviet Union, and the Russian military kept it in service on the Baltic Sea for a number years under the name MS *Podvizni*. When the ship was irreparably damaged in an accident, the Soviet Navy towed it north and anchored it off Maly. For reasons of secrecy it was renamed *Kit*, or whale.

Alexander Kukushkin, one of the men who took part in the nuclear tests, later told of how troops stationed on Maly filled the vessel with various animals. The upper and lower decks were crammed with rabbit and rat cages, and dogs were tied up there too. Then explosives were detonated on the decks and in the bilge. "We used shells three hundred millimeters in diameter," Kukushkin recalled. "The shells were filled with chemical explosives and each contained an ampule of uranium 235. The uranium was dispersed by the force of the blast. The entire area was contaminated in this fashion." Kukushkin described how the experiments put various types of protective clothing and anti-nuclear materials to the test. He also reported that the unit commander, a man named Dvorovoi, died in 1954 of radiation poisoning. "Two others had already died of the same causes at the test site," Kukushkin remembered. "Back then we knew very little about the effects of nuclear radiation."

After bombs had been detonated, officers ordered their underlings to go to Ground Zero and take water samples with their bare hands. Several sailors collected large salmon whose airbladders had been torn apart by the force of the explosion and roasted the fish over campfires. We do not know how many of the participants in these tests eventually died of exposure to radiation.

It was hardly exceptional for soldiers themselves to be used as guinea pigs. As part of one of their most devastating atomic tests, the Soviet leadership ordered more than 40,000 Red Army soldiers to march to Ground Zero immediately after a mushroom cloud had dissipated. This test took place in Tozk in the Southern Urals, a region of woodland hills and valleys similar to Germany, the country both NATO and Warsaw Pact signatories had nominated as the primary battlefield, should it ever come to World War III. In the planning stages of the test, the Samara River was even referred to as the Rhine. Deputy Soviet Defense Minister Georgy Zhukov noted his satisfaction with the realistic conditions of the exercise, which he witnessed from the safety of a bunker together with representatives of Communist "brother countries." While they chatted away, talking shop, soldiers outside were being exposed to radiation without any protection at all. Sometime later, Zhukov went out to inspect Ground Zero for himself— inside a tank lined with thick layers of lead. As was the case with Lake Ladoga, the truth about the Tozk atomic test only emerged after the end of Communism, and these weren't the only dirty secrets the Soviet military kept during the four decades of the Cold War.

There is no reliable information about what happened to the *Kit* after the series of tests was completed. According to one source, the Soviet navy continued to use the vessel up until 1961, whereas the newspaper *Izvestia* insists that the ship was decommissioned in 1954. However long the Soviets may have used the German vessel for experiments, at some point they lost interest and sank the *Kit* together with her radioactive cargo. Photographs allegedly taken in 1978 show a gigantic ship, rusted through and decaying, lying half-submerged in a bay.

In response to pressure from the media, Russia's nuclear authority decided to examine the case in the early 1990s. Measurements taken on board the *Kit* yielded devastating results. The hulk was thoroughly contaminated—even the peeling paint was "hot," i.e. massively radioactive. Radioactive sludge was also leaking from the interior of the minelayer into Lake Ladoga. Shocked, Anatoly P. Sklyanin—the chief of what was then St. Petersburg's nuclear safety department—told a journalist from the *Los Angeles Times*: "We cannot leave for our children and grandchildren these dark stains that our parents were forced to create." A short time later, the holes in the *Kit* were plugged, wherever possible, with two artificial chemical compounds. In the summer of 1991, the St. Petersburg nuclear safety authority secretly had the radioactive ghost ship towed from Lake Ladoga to the island of Novaya Zemlya thousands of kilometers farther north. There the trail of the *Kit* goes cold. In all likelihood, the ship was sunk once again and now rests at the bottom of the Arctic Sea.

Even today, almost sixty years after the Soviet Union

carried out atomic tests in Europe's largest interior lake, much of what happened remains shrouded in darkness. One of the most intriguing eyewitness reports maintained that an underground atomic bomb was detonated on the border with Finland. According to this testimony, the Russian Navy constructed a scale model of Manhattan on the shore of the lake to test whether the city could be flooded by an artificially induced tsunami. But the story may by apocryphal. A 1992 report drawn up by the American Department of Energy, entitled "Radioactive Contamination of the Arctic Region, Baltic Sea, and the Sea of Japan from Activities in the Former Soviet Union," makes no mention of any such test.

That DoE report provides an overview of the various nuclear legacies left behind in the extreme north of the planet. Along with the *Kit*, massive amounts of atomic waste were dumped in the sea around Novaya Zemlya. The DoE lists the following radioactive remnants of the Soviet era:

1. The Novaya Zemlya deep-sea trench—a cargo vessel with a damaged (submarine) reactor (1,700 curies [reported to be 170,000 curies by *Rossiyskaya Gazeta*, June 23, 1992]), 1,450 submerged containers with radioactive waste, and a tanker with liquid radioactive waste

2. Neupokoyeva Bay—solid radioactive waste with an overall radioactivity of 3,400 curies

3. Tsivolki Bay—4,750 containers of radioactive wastes, the lighter *N. Bauman*, the midsection of the ice-breaker *Lenin* with three damaged reactors and crane assembly

4. Oga Bay—850 containers of radioactive wastes

5. Stepovogc Bay—1,850 containers of radioactive wastes and a damaged nuclear submarine containing two reactors with nuclear fuel still inside (sunk at a depth of 35–50 meters [*Rossiyskaya Gazeta*, June 23, 1992])

6. Abrosimov Bay—550 containers of radioactive wastes and sections of four damaged nuclear submarines (a total of eight reactors, three of which contain nuclear fuel)

7. Blagopoluchiya Bay—650 containers of radioactive wastes

8. Techeniy Bay—a damaged reactor (without nuclear fuel) with an aggregate activity level of 1,856 curies

9. Open sea—650 containers of radioactive wastes

The list is by no means comprehensive. We can safely assume that a lot more covertly dumped nuclear waste has yet to be discovered. Moreover, the list does not include the disposal of nuclear waste in and around the base for the Soviet Navy's North Sea fleet in Murmansk. Nor is the disposal of radioactive material the only problem in Novaya Zemlya. The Soviet military also conducted atomic tests there as late as 1990. More than 220 nuclear explosions have been reported in the region.

The focal, organizational point of these top secret Soviet military operations was a military settlement in the southern half of the Novaya Zemlya archipelago. It can be reached by plane from the Soviet polar metropolis of Arkhangelsk, but every time a flight departed for or arrived from there, the announcement board at the airport would show the cryptic "Amderma-2." In official government jargon, the settlement, which was not shown on any maps, was known as Arkhangelsk 56. In reality, it was located in Belushya Guba on the southern island of Novaya Zemlya. But even if you make the long trip to this remote settlement, you'll still be a long way from the actual test sites. Novaya Zemlya is twice the size of Denmark, or roughly the size of South Carolina.

Three hundred kilometers farther north, along the Matochkin Strait that divides the north and south islands, was another, much smaller military settlement. It was there, in Sukhoy Nos Zone C, on October 20, 1961, that the Red Army detonated what was by far the biggest atom bomb in human history. The AN602 or "Tsar Bomba" weighed twenty-seven tons and had a yield of fifty megatons, making it 1,400 times as powerful as the bombs dropped on Hiroshima and Nagasaki together. The resulting explosion could have easily destroyed Paris and its greater metropolitan area. The seismic shock wave it unleashed traversed the globe three times and was measured as far away as New Zealand. In Norway, it shattered window panes. Presumably not even the Russian military could explain what the point of developing such a doomsday weapon was.

The negative impact of the Soviet Union's nuclear testing mania on Novaya Zemlya has been kept secret or downplayed. Even today, some Russian scientists maintain that the atomic aftereffects aren't really all that bad. In 2004, one Russian academic even proposed in an International Atomic Energy Agency (IAEA) report that the psychological effects of coming from a region used for nuclear testing were worse than the dangers of contamination and fallout. The stress of "being forced to relocate," the author opined, could take up eight years off an individual's life.

Large parts of the gigantic archipelago remain uncontaminated—no one disputes that fact. But there are enormous radioactive hot-spots in and around the test sites, and it was only in 1990 that details about them became public. The first bits of information came not from official government sources, but as a result of the activities of environmentalists. In the 2004 IAEA report, two Russian scientists write in tones approaching outrage of the "illegal penetration of Greenpeace activists" into test zone B. The "MV *Greenpeace*" was pursued by the KGB-operated icebreaker *26th Party Conference* and a smaller vessel. When called upon by radio to specify their intentions, the skipper of the Greenpeace boat allegedly answered: "A nuclear-free world." After a chase, Russian authorities boarded the MV *Greenpeace*, but not before four activists succeeded in landing on Novaya Zemlya in rubber dinghies. They set off with Geiger counters. The levels of radioactivity they measured on the coast weren't all that high, but as they proceeded into the hills inland, radiation levels rose. An article in the weekly German newspaper *Die Zeit* described what happened next on this unique expedition:

> Three kilometers inland, the four activists discovered an entrance to a tunnel. Usually, despite their poor reputation, nuclear facilities are clinically precise and, technically speaking, extremely efficient. But the tunnel looked like a long-abandoned mine shaft. Rusty generators were lying around amidst tipped-over trolleys, decrepit Quonset huts, twisted section steel and cut-open rolls of cable. It was chaos.

The Geiger counters went crazy, going off the scale for over an hour. According to *Die Zeit*, after eleven hours on the island, the members of the expedition had been exposed to radiation levels far above the norm even for professional radiologists.

In the Soviet era, the intruders would have been thrown in jail for espionage. Thanks to Gorbachev's perestroika, the four Greenpeace activists were detained and then released a short time later. They had succeeded in exposing a problem whose potentially disastrous consequences made it impossible to ignore.

Most of the contaminated Novaya Zemlya archipelago is only sparsely settled, but other Soviet nuclear test sites were located near far more densely populated areas. For residents of such regions, this chapter of the Cold War is anything but complete. The inhabitants of the Kazakh village of Sarzhal, for instance, are among 200,000 people who live in the immediate vicinity of the Semipalatinsk test site, an area the size of Wales, which the Soviets also called "the Polygon." In 2010, an al-Jazeera film crew traveled there and recorded some of the residents' memories. Villagers told how in the 1950s Soviet officials had appeared and ordered them to lie down on the ground of the steppe and shut their eyes. A short time later, there was a bright flash, and a second sun ascended in the sky. The villagers had no reason to suspect that the explosion might put their health at risk; they were told nothing about the possibility of radiation. After the explosion, men in protective suits and gas masks arrived and held measuring instruments up to the entire village population's chests. One young man, a fellow named Amanbek Kazenov, was taken away to a secret laboratory, where doctors subjected him and others to days of tests. "We were their guinea pigs," he says today. The facility was ostensibly a clinic for treating brucellosis, but this was a smokescreen. In reality, the KGB used it to investigate the effects of radiation on the human body.

The Red Army viewed Kazakhstan as something approaching a Russian colony, treating the inhabitants of the Polygon with cavalier disregard. Moreover, Kazakhs weren't the only ethnic-cultural group exposed to radiation. Some 600,000 ethnic Germans whom Stalin had deported to the steppes in 1941 were subjected to similar treatment. One of them, a man named Wilhelm Leier, was taken to the town of Semyonovka, only eighty kilometers from the Polygon. He told Russian journalist Igor Trutanov that after explosions, when it rained in his village, people would suffer skin burns. About ten years later, a number of children died of mysterious diseases. Leier remembered the USSR's first hydrogen bomb. "After the detonation of the super-bomb in August 1953, there was a storm that shattered the windows in all the houses in the village," he recalled. "A lot of people were injured by shards of glass. You can hardly imagine it. A second sun rose in the sky, and most people ran out of their houses, although some watched from inside through their windows. A few minutes later, an unexpected wave of pressure smashed the panes of glass. Luckily, no one went blind ... years later, I learned that we had witnessed the first hydrogen-bomb test."

Today the aftereffects of radiation are evident throughout the Semipalatinsk. Oncologist Arumert Kim reports that cancer rates in the vicinity of the Polygon are 2.3 to 3 times higher than in the general population. The deadly tumors also appear earlier in life, so that cancer sufferers in this area are noticeably young. The fallout knows no generational boundaries. There are families in Sarzhal in which every single child was born with serious disabilities. No one told the villagers that the nuclear tests carried out in their backyard might damage their DNA.

Similar stories of human suffering recur in other regions where atomic tests are known to have been staged as well as in areas where the Soviet Union used to produce nuclear weapons. In 1957, there was a level-6 explosion at a nuclear fuel reprocessing plant in a closed city known as Chelyabinsk-40 in the Eastern Urals. It contaminated the entire surrendering area. Twenty-two villages had to be evacuated. It is now considered the third-worst nuclear disaster in history, behind Chernobyl and Fukushima. But it wasn't until the collapse of the Soviet Union that local inhabitants or the world at large were fully informed about the seriousness of the incident.

We should be cautious about seeing such indifference to the fates of individuals as an exclusive feature of the now-defunct Soviet Union. Other nuclear powers have displayed remarkable irresponsibility and cynicism in carrying out atomic testing. The permanent contamination of radioactive wastelands like Novaya Zemlya and the declarations of the Semipalatinsk as "uninhabited" were hardly Communist exceptions to the human norm. On the contrary, as we will see, variations on these stories are heard elsewhere.

Moreover, the biggest current problem with the USSR's nuclear legacy is one that only emerged in the post-Soviet era. In 1990s Russia, the entire social order collapsed, leaving

a power vacuum. The old organs of the Soviet state still oper-
ated, but they were spinning their wheels, while nebulous
mafia and criminal networks gained influence throughout
the country. A healthy market arose for illegally traded nuclear materials. Weapons-grade uranium and plutonium was
sold back and forth throughout Europe. In the city of Murmansk, three uranium fuel rods went missing and ended up
in the hands of smugglers. In 1994, 900 milligrams of highly
enriched uranium suddenly turned up in the German town
of Landshut. That same year, Czech police pulled off a sensational coup, confiscating three kilos of top-grade nuclear
bomb-making material in Prague.

It would also be a mistake to dismiss these episodes as
anomalies from the "Wild East" period. While the most egre-
gious holes in Russia's nuclear economy were plugged in the
course of the 1990s, the biggest problem remained. Until fairly
recently, anyone could procure unlimited amounts of nuclear
material—completely risk-free. All you needed was a solid
pair of hiking boots.

Between 1949 and 1989, the Soviet Union detonated 496
atomic weapons in Semipalatinsk. In the 1950s, there was
barely any break between explosions, and today's experts have
been left searching for solutions to deal with the aftereffects.
Even in modern nuclear weapons, only a small proportion
of the radioactive material inside a bomb explodes when the
warhead is detonated. The rest gets scattered in every direction. Semipalatinsk was the site of aboveground tests in which
a maximum of 30 percent of the plutonium and uranium
contained within the bomb was consumed. The remainder
of the weapons-grade radioactive material—usually several
kilos—was often just left lying around the steppes. Someone trying to acquire uranium 235 and plutonium 239 for

nefarious purposes would only have to travel to Kazakhstan and go hunting along the grounds of the Polygon. After the demise of the USSR, the Red Army was in such a hurry to evacuate its nuclear test site in Kazakhstan that it even forgot an atomic bomb in one of the underground shafts there. It wasn't until 1995 that the device was defused and disposed of. In the aftermath of that incident, a Kazakh military unit was assigned to patrol the test site, but it consists of only 500 men, a hopelessly small number to keep tabs on an area that measures 18,500 square kilometers.

The massive heat generated by the nuclear explosions fused some radioactive substances with the surrounding stone. It would be difficult for someone who collected the glasslike, radioactive fragments to extract the material in order to make weapons. But it would not be impossible—as an initiative of the U.S. government shows. After September 11, 2001, alarm bells were ringing all over in Washington, and a secret meeting was arranged with representatives from Russia and Kazakhstan at which political leaders agreed on "Operation Groundhog." It had Kazakh laborers covering particularly contaminated ground within the steppes with concrete, while nuclear experts sealed the underground shafts that had been left behind from untold numbers of nuclear tests. The three governments thought they had the problem under control. But one of the diplomatic cables published by the whistleblower website WikiLeaks in 2010 contained a reference to Operation Groundhog. In it, former U.S. Ambassador to Kazakhstan Richard Hoagland called the securing of weapons-quality material within the former Soviet test site Semipalatinsk the most important aspect of the Cooperative Threat Reduction program for destroying weapons of mass destruction in the former Soviet Union. The cable made it clear that Operation

Groundhog was not only still underway, but was being expanded. That provoked the question: Why was this necessary? To follow statements made by one high-ranking NATO official, the reason was that no one could say where the radioactive hot-spots were or how many of them existed. Presumably, the more experts investigated the problem, the more it grew. But the findings of the decontamination teams associated with the operation in Kazakhstan remain classified.

What we do know is that even after the first phase of the cleanup operations, people could still go in and out of the gigantic test site at will. Shepherds, for example, took their sheep to graze on the steppes of the Polygon and made soup from the bones of radioactively contaminated horses. Junk dealers also actively scavenged in the test grounds. In times of high prices for raw materials, there were plenty of profits to be made recycling potentially radioactive metal. The Chinese were particularly eager to acquire salvaged copper, which was simply lying around within the Polygon, waiting to be picked up. A number of particularly enterprising gangs even blew the concrete seals off the bomb shafts to get at the copper wiring used to detonate nuclear weapons. At least ten scrap-metal dealers died of radiation poisoning. The contaminated metal was melted down, sold off, and used elsewhere. It is hardly beyond the realm of the imagination that similar raids may have been carried out by people with far less benign motivations. It's hard to conceive that a terrorist organization like al-Qaeda would be able to construct a nuclear bomb from such material. But it would be far easier to combine radioactive material and conventional explosives into a dirty bomb. It's an unsettling scenario.

The Polygon was a graveyard not just for radioactive detritus, but for millions of U.S. dollars. Operation Groundhog was originally conceived as a cheap cleanup measure—but the program now costs as much as 50 million dollars annually. The cost of securing radioactive material is only part of the expense. Securing the entire site required even more financial resources. Kazakh troops had to be upgraded, fences built, and the steppe equipped with motion detectors. Unmanned drones, sent by the United States Defense Department, now patrol the skies over the Polygon. No one knows when, if ever, Operation Groundhog will be completed, or how dangerous the nuclear remnants on the site remain today.

CHAPTER THREE

The Myth of Tactical Nuclear War

John Wyndham's classic 1951 apocalyptic science fiction novel *The Day of the Triffids* starts with what might be described as a dirty trick played upon humanity. One night, a series of beautiful lights, similar to fireworks or lightning, appears in the heavens. Everywhere across the globe, people congregate to watch the awesome spectacle. London's Trafalgar Square is packed with crowds, and hundreds of people squeeze their way onto rooftops with telescopes to better view the fiery night sky. No one can take their eyes off the show. The masses even break out into cheers after particularly sensational light effects. The following morning, humanity is in for a nasty surprise. Everyone who looked at the lights in the skies has gone blind. Once more the streets fill up. Only this time the people outside are not thrill-seekers, but desperate, panicking souls upon whom eternal darkness has descended. Wyndham never explains what the lights were or the cause of their devastating effect. Perhaps, the novel's narrator surmises at one point, they were a newfangled weapon accidentally detonated in space.

Seven years after Wyndham's novel was published, reality caught up with apocalyptic fiction. On the evening of August 1, 1958, people in Honolulu, Hawaii were witnesses to an artificial heat lightning like that in Wyndham's novel. The light show commenced without warning. When the phenomenon recurred in the summer of 1962, bar owners in the Hawaiian capital put up signs promising that their terraces had the best view in the city. Thrill-seekers followed the heavenly spectacle with a mixture of enthusiasm and awe, but it too had negative side effects. Radio signals were disrupted, and a major telephone network broke down. Burglar alarms suddenly sounded everywhere, and some 300 streetlights went out. A number of satellites could no longer send signals back to earth from their orbits. Years later, it turned out that some of these satellites, including the Soviet Kosmos V and the first American TV satellite, Telstar I, had been damaged. Three were rendered completely non-operational.

What caused these mysterious phenomena were American atomic tests in the stratosphere. Balloons and rockets transported nuclear warheads fifty kilometers above the earth's surface, where they were detonated. In a test called "Starfish Prime," a Thor rocket even took a bomb up to an elevation of 400 kilometers in the thermosphere. No one went blind as a result of the detonations, although there were some fears of that happening. Radiation released by the nuclear explosions could have scorched human retinas. As part of tests in the earlier "Hardtack" series, rabbits given the window seats on flights sent to monitor the zero points of detonations all suffered damage to their eyesight. During the "Bluegill Triple Prime" test in October 1962, there was an accident. Two soldiers manning an advanced surveillance looked up at the sky without protective goggles and later complained about disrupted vision. One of them, a naval officer, never regained his full eyesight and had to be discharged from service.

It was thus no accident that the high-altitude nuclear tests were carried out above a remote region in the Pacific, high above Johnston Atoll. Distance was the reason people could watch them safely in Honolulu. The experiments were very colorful—a fact reflected in code names like "Bluegill," "Teak," and "Orange." Some failed and produced only a tiny "fizz." Others had a yield of more than three megatons and spread pulsating light across the heavens. The Soviets also carried out high-altitude nuclear tests and were always curious about what their rivals were up to. Every time America sent a rocket with a nuclear payload into space, Soviet research ships showed up as uninvited guests to monitor the results.

The tests were aimed at answering a variety of questions. First and foremost, the U.S. military wanted to know whether nuclear explosions in the earth's atmosphere were capable of destroying enemy ballistic missiles—an attractive proposition in an age when the world's superpowers were arming themselves to the teeth. America also wanted to test the effect of atomic detonations on enemy communications and electronic capabilities. Nuclear explosions in the atmosphere create a strong electromagnetic pulse (EMP) that was devastatingly effective at disabling radio installations and electronic devices. Washington also wanted to know whether nuclear warheads could be used as weapons against enemy satellites and spaceships. Scientists discovered that the bombs were far *too* effective. The detonations created belts of radiation that encircled the earth for months, affecting both friends and foes equally. NASA considered the tests a potential threat to America's civilian space program. No one wants to send U.S. astronauts through a belt of radioactivity, and in a personal audience with John F. Kennedy, NASA head James E. Webb demanded an immediate end to the tests. As a result, the "Uracca" experiment, planned for an altitude of 1,300 kilometers, was cancelled. Stratospheric nuclear weapons had proved to be of no military value anyway. The risk was simply too great that such weapons would disable one's own satellites and other facilities, to say nothing of harming soldiers and civilians.

Fortunately, the atmospheric tests did not have all the effects scientists imagined they might. The tests did not lead to measurable changes in the earth's climate. There were neither catastrophic storms nor drastic rises and drops in temperatures. Another major fear also proved unfounded. Some scientists had warned that stratospheric tests could tear a hole in the ozone layer, allowing harmful UV rays to penetrate to the earth's surface. But the ozone layer remained intact.

Nonetheless, the tests did lead to the radioactive contamination of the sensitive mantle protecting the earth. By the late 1950s and the early 1960s, radioactivity was on the rise globally, hitting levels not even reached after the Chernobyl catastrophe. Over time, the radioactivity released into the upper reaches of the atmosphere by the high-altitude tests settled down into the lower atmosphere. We can only guess what effects the accompanying increased radiation levels had for human beings and their environment.

For people in the South Pacific, however, the tests that made up "Operation Crossroads" were far more devastating than the Starfishes, Teaks, and Bluegills. The sad destinies of those directly affected by the 1948 tests at the Bikini Atoll, as well as later ones at Eniwetok and other atolls, have been well documented. Archive film footage shows natives of Bikini lugging their scarce belongings onto a navy transport ship that will take them to Rongerik, an uninhabited and—as it would turn out—uninhabitable island. This was the beginning of a decades-long odyssey to find the islanders a new home. There was no returning to Bikini. The island remains contaminated today. Currently some of the islanders live on Majuro in the Marshall Islands, while others exist in exile in Arkansas.

The U.S. military told the Bikini Islanders that their homelands would be sacrificed in the name of world peace, but in reality America's South Pacific nuclear tests were motivated by somewhat less noble concerns. The tests were primarily intended as demonstrations of strength, showing the rest of the world, and the Soviet Union in particular, just how mighty the U.S. atomic arsenal was. To this end, the tests "Able" and "Baker" were filmed by dozens of cameras mounted on ships and airplanes. The U.S. military was also interested in investigating whether nuclear weapons had made warships obsolete. Maintaining a navy, the logic ran, would be senseless if the Soviet Union could destroy a whole fleet with a single bomb. If so, Washington could save the money it invested in expensive warships. As grotesquely Cold War–bound as this logic may appear today, the U.S. Navy did in fact tow in a whole fleet of decommissioned warships to be used in the nuclear tests. The vessels included an outdated American aircraft carrier, the captured German heavy cruiser *Prinz Eugen*, and the *Nagato*, the Japanese destroyer on board which the orders to attack Pearl Harbor had been issued. They were anchored just off the Bikini Atoll.

As recorded on film, the test itself was a macabre, bizarrely orchestrated spectacle. Before the first explosion, a famous tenor performed a stylized song of farewell to this South Seas paradise and its soon-to-be-extinct culture. The chieftain of the Bikini islanders, known as "King Juda," had been given a front-row seat on the deck of one of the ships not earmarked for destruction. Through dark protective goggles he watched on, hardly flinching, the very picture of nobility, voluntarily sacrificing his home island to a supposed historical necessity that would benefit all of humanity.

Much to the relief of the U.S. Navy leadership, the

twenty-three-kiloton nuclear bomb Able did not sink the majority of the test ships. Most remained intact, albeit radioactively contaminated. The second bomb, Baker, was detonated underwater. A huge pillar of water shot skyward, tossing a number of warships through the air like toys. Nonetheless, for all the destructive force of the bomb, it still took hours before the *Nagato* and nine other target ships sank. The rest of the target fleet survived, although badly damaged. Thus the Baker test, too, failed to provide a clear answer to the question posed by the experiment. It seemed impossible to destroy a whole naval fleet by pressing a button. On the other hand, fleets were hardly immune to nuclear attacks.

The irony is that neither side in the Cold War seems to have considered that attacks on ships would have a decisive role in an all-out nuclear war. That, however, did not prevent the military leadership on both sides from giving free rein to their imaginations. Particularly in the 1950s and 1960s, the U.S. Defense Department devoted considerable attention to the development of "tactical" nuclear weapons. In contrast to strategic weapons, which were aimed at deterring and, if necessary, destroying the enemy, tactical weapons were intended to achieve specific military aims on a nuclear battlefield. Tactical nuclear bombs had less range than strategic ones, although not necessarily less explosive force. They included naval bombs designed to sink enemy ships and submarines as well as atomic grenades to be deployed against columns of approaching hostile forces. Many tactical atomic bombs did without delivery systems and served solely defensive purposes. In Western Europe, for instance, there were units of soldiers whose only mission was to destroy major streets in case of war. The Brenner Pass between Austria and Italy was mined with

explosive devices that could have demolished the raised highway between the two countries. In West Germany, the Second
Brigade of the Third U.S. Armored Division was charged, in
case of a Soviet invasion, with plugging the "Fulda Gap," a
passage through the central German mountains that the Red
Army could have theoretically used for a lightning advance on
Frankfurt am Main. The U.S. troops were tasked with slowing
the invasion by detonating atomic mines as part of a so-called
"Zebra package." The idea was to close narrow gaps in the
landscape, for instance, by creating artificial landslides.

Nuclear landmines were known as Atomic Demolition
Munitions or ADMs, and tunnels and mine shafts were dug
throughout West Germany to house these explosives. Nonetheless, ADMs made no more sense than nuclear detonations
in space. Conventional explosives would have been just as effective for demolishing bridges and major passageways—and
would not have had the terrible radioactive aftereffects. In
reality, ADMs were another way for NATO to bare its atomic
teeth, suggesting that it could halt a Soviet invasion of Western Europe without having to deploy its full nuclear arsenal.
In that respect ADMs were like strategic nuclear bombs, the
difference being that the former were intended to have a deterring effect *after* war had already commenced. ADMs have thus
been rightly described as purely political weapons.

It is highly doubtful whether these and other tactical
atomic weapons could have "contained" a nuclear war and
thus saved the human race from extinction. In fact, they
played only a limited role in the strategies of the USSR. For
the Soviets, nuclear war was nuclear war—regardless of the
means by which it was conducted. The military leadership of
the Red Army thus saw tactical nuclear weapons as but a part

of the arsenal that would be used, should it come to a full-scale confrontation between the superpowers.

For its part, NATO should have been aware of how specious the distinction was between tactical and strategic nuclear arms. In a 1957 exercise, America and its allies simulated the course of a nuclear war in Europe, in which 355 tactical nuclear weapons were used. The resulting casualties of even such a "limited" war were calculated at 1.7 million people. With those numbers of dead, the only things limited about the war would have been the lack of destruction in North America. In any case, it's hard to imagine the Soviets accepting losses on that scale. In the 1990s, a former member of the Soviet General Staff, Adrian Danilevich, confirmed that the USSR would have answered any nuclear attack by the West, even with only a single bomb, with a massive atomic counterstrike.

Nonetheless, the American pipe dream of a limited nuclear conflict, in which neither side would employ their complete atomic arsenal, was astonishingly persistent and led to bizarre and dangerous inventions, including *the* nightmare weapon of the Cold War. The neutron bomb was a tactical enhanced-radiation weapon designed to emit massive amounts of radioactive neutrons. Fast neutrons can only be absorbed by substances containing hydrogen, so theoretically they would have penetrated steel armor and reinforced concrete. It was thought that tank divisions would have no chance against such a weapon, the most inhuman of all nuclear warheads. Humans exposed to such radiation would live on for days before suffering extremely painful deaths.

Other Cold War weapon innovations were nearly as terrible. The DASH helicopter drone was intended to drop nuclear bombs into the ocean in order to sink submarines. The vehicle was so prone to crashing that even the U.S. Navy was wary of it. Although the drones were deployed in the Vietnam War, they were only used for surveillance purposes. Thankfully, they were never armed with nuclear depth charges.

The absolute pinnacle of senselessness, however, was a tiny nuclear weapon nicknamed the "Davy Crockett." It consisted of a small, mounted, recoil-less gun that fired a thirty-four-kilogram nuclear projectile. As a weapon, it was extraordinarily imprecise, but its inexactitude was compensated for by its strong radioactivity, which would have killed every living thing within 400 meters of impact. In the confusion of ground battle, the weapon was more likely to hit friendly troops than any hostile combatants. The Davy Crockett was tested for the first time in Nevada in 1962, and in the years that followed, thousands of them were produced and deployed around the world, particularly in West Germany and South Korea. The small size of the weapon and its explosive capacity of only ten to twenty tons of TNT concealed how extraordinarily dangerous the system actually was. Military handbooks heaped scorn on such weapons, and a United Nations study from the year 2000 concluded that tactical nuclear weapons (TNWs) represented a source of extreme peril because of their physical form, the locations in which they were deployed, and the instructions soldiers were given on how to use them. The compact size and the lack of electronic safety catches on older models meant that they were more prone to being stolen and misused than other nuclear weapons. The report went on to enumerate other security risks:

1. The intended use of TNWs in battlefield and theatre-level operations in conjunction with conventional forces encourages their forward basing, especially in times of crisis, and in certain situations movement of

TNWs might actually provoke a pre-emptive strike by the other side instead of deterring it; and

2. An orientation towards the employment of TNWs in conjunction with conventional forces and a concern about their survivability argues for the pre-delegation of launch authority to lower level commanders in the theatre, especially once hostilities commence. This might result in diminished control by the political leadership over TNWs.

For those reasons, the authors of the UN study concluded that "the very existence of TNWs in national arsenals increases the risk of proliferation and reduces the nuclear threshold, making the nuclear balance less stable." In other words, a nameless field commander could have inadvertently unleashed an all-out nuclear conflict, bypassing the complex command structures that involved the governments and heads of state of the United States and the Soviet Union. In producing thousands of such tactical nuclear weapons, both sides accepted what in retrospect seems like an irresponsible, indeed insane risk.

The miniaturization of nuclear weapons also had another consequence: continual rumors about briefcase- or backpack-sized terrorist bombs capable of being smuggled into cities like New York or London and detonated there. Frenzies of this sort of speculation recurred at regular intervals, often after information provided by dubious characters like the late Russian general Alexander Lebed. On one occasion, the officer and politician claimed that the KGB had constructed more than one hundred "backpack bombs." Sometime later, he asserted that most of them had been stolen. With Lebed, it's difficult to separate the facts from the self-aggrandizing fictions. While the

Cold War did see the creation of atomic mines small enough to strap to the backs of paratroopers, a complex military infrastructure was needed to actually use them. One problem with miniature bombs was that radiation from nuclear material tended to damage the complex electronic mechanisms needed for the detonators. Stealing weapons like this would have been idiotic. Without the proper maintenance, they wouldn't have functioned. Significantly, none of the suitcase or backpack bombs that Lebed claimed had disappeared was ever located.

Not so long ago, the idea of dwarf nuclear weapons underwent a sudden renaissance. Early in his first term of office, then U.S. President George W. Bush promoted the development of so-called "mini-nukes." What he meant were bunker-busting bombs capable of penetrating far below the earth's surface. But Bush encountered immediate resistance to the idea, and in 2005 he withdrew his suggestion that these sorts of projects be funded. Whether or not such weapons are being secretly developed is anybody's guess.

That's another reason to suspect that talk about tactical nuclear weapons is primarily a means of exerting political pressure on adversaries. Former U.S. Secretary of Defense Donald Rumsfeld made no bones about America's willingness to employ mini-nukes as preemptive weapons. Nonetheless, in order to destroy an underground bunker, a "miniature" atomic bomb would have to have a yield of at least 300 kilotons of explosive force. The bomb dropped on Hiroshima, which killed more than 100,000 people, had a force of "only" thirteen kilotons.

CHAPTER FOUR

The Radioactive Cowboy, or How Alaska Got the Bomb

Susan Hayward had rotten luck with men. The Hollywood diva may have played "tough woman" roles, such as Empress Messalina in the sword-and-sandal epic *Demetrius and the Gladiators*, but in real life she was insecure and moody. She was married to Jess Barker, one of those typical square-jawed B-movie actors and a ladies' man who cheated on her at every opportunity. When his career began to go downhill, he tried to borrow a large sum of money from his wife to start an oil company. Hayward refused, and it came to a fight in which she bit her husband on the arm. He then allegedly dragged her through the backyard, naked, and chucked her into their swimming pool.

Hayward fled from Barker into the arms of a man whose eccentricities were legendary even in a country that prides itself on having no shortage of oddballs. Howard Hughes was worth billions, having turned a tool-making company he inherited from his farther into a diverse and flourishing corporation. The billionaire loved golf, airplanes, and Hollywood films. But he had little success when he tried his hand as a movie producer and a studio boss. Hughes's productions did, however, bring him into contact with the most beautiful women the dream factory had to offer, and he wooed them with almost suffocating attention. Hughes was not only

generous, but jealous. He kept his lovers, including Ava Gardner and Terry Moore, under twenty-four-hour surveillance.

Hughes had been friends with Hayward for years before
they got to know each other much better on the set of *The
Lusty Men*. The result was merely a casual fling, but with Hayward, too, Hughes had trouble letting go. To maintain his
influence over her, he forced her to take on the lead role in the
most dubious movie, in terms of taste, that he ever produced.
The script for *The Conqueror*, a costume film about Genghis
Khan, was completely wretched. The story focused on the
Mongol warrior's youth, a period in which he was obsessed,
in the best tradition of the Hollywood Western, with aveng-
ing his father's death. Along the way, young Genghis wins the
heart of shrewish Tartar Princess Bortai, melodramatically
swearing his undying love to her. Against an exotic backdrop,
the film runs through one 1950s macho and sexist cliché after
another. Hayward struggled to maintain some dignity as the
female lead amidst all of Hughes's intellectual flotsam and
jetsam, even though as a red-haired Anglo-Saxon she hardly
made a credible Tartar princess. The script reduced her to a
prop, someone as utterly dependent on Genghis as Hughes
always wished women were on him in real life. Playing the
passionate lover and fearless warrior Genghis was none other
than John Wayne. Allegedly, the veteran gunslinger had prac-
tically begged for the utterly incongruous part.

It's hard to imagine two more thoroughly miscast actors.
Wayne, who was far more at home playing upstanding sheriffs,
looked ridiculous in his fake Genghis beard, and Hayward,
who was supposed to be an ill-treated prisoner in a dungeon,
was thoroughly glamorous and always immaculately made
up. Even during the action scenes, not a single lock of her
hair ever fell out of place. In contrast to Wayne, Hayward had

been against the project from the very beginning. But Hughes had gotten his way, wearing her down with a combination of threats and money.

As a setting, the filmmakers had chosen Snow Canyon in southwest Utah. The landscape may have been nothing like the Mongolian steppes, but that didn't matter to director Dick Powell. The canyon was near a small town called St. George, and it was from there that, a short time previously, prospectors had headed out into the desert armed with Geiger counters. When they reached Snow Canyon, the measuring devices went crazy, yet upon digging, the prospectors didn't find any uranium. Only a year before Powell and his crew began shooting, the U.S. Army had carried out a test code-named "Harry" in neighboring Nevada. It would go down as one of the dirtiest atmospheric detonations of all time. From a military perspective, the test had been a rousing success. About 1,000 soldiers had been able to view Harry's fireball for a full seventeen seconds. The bomb contained relatively little nuclear material, but its destructive force had been devastating—especially for the environment. At least a third of all radiation released by all American nuclear tests between 1951 and 1958 came from this one exercise. The pillar of smoke had ascended to an elevation of 11,600 feet before topping out and heading east. A radioactive cloud drifted off toward St. George, 270 kilometers away from Ground Zero. For several hours, unsuspecting residents were showered with radiation. Many of the 5,000 "downwinders" who populated the town would later develop cancer—a late result of Harry's fallout. Yet the authorities rejected the idea that the high levels of radioactivity in Utah could have anything to do with atomic tests in Nevada.

Nor was St. George the only place that had been contaminated. In Snow Canyon, radioactive particles from the main

nuclear cloud had collected as if in a funnel. It was precisely
there that, at Hughes and Powell's insistence, the first clap-
per board was snapped shut for the filming of *The Conqueror*.
The producer and at least part of the cast and crew must have
been aware of the radioactivity—John Wayne even posed for
a photo with a Geiger counter in Snow Canyon—and there
is absolutely no doubt that Hughes knew about the potentially lethal effects of radiation exposure. He had co-financed
a 1953 film, *Split Second*, that had been set in a nuclear testing
range. The director of that film was none other than Dick
Powell. Nonetheless, the two men sent their team out into the
gleaming sands of the valley. Five thousand Indian extras were
brought in for the monumental battle scenes, 1,000 horses gal-
loped across the 120-degree heat of the landscape, and dozens
of aspriring starlets hired to play concubines strolled around
half-naked, fulfilling male fantasies.

Shooting got off to an inauspicious start. John Wayne hurt
himself in a riding accident, and Susan Hayward was attacked
by a "tame" panther. When Genghis's falcon also got sick,
work on the film was suspended for days. Meanwhile, Wayne's
wife and Hayward nearly came to blows after the actress supposedly fell in love with the rapidly graying Duke. Amidst all
the accidents and misunderstandings, though, the actors were
working in radioactive filth. Huge fans threw up sand and dust
to heighten the drama of the battle scenes. At the end of each
day, the actors had to be cleaned off with air presses and hoses.
Stuntmen who threw themselves from their horses swallowed
mouthfuls of sand. Moreover, after the on-location work was
finished, Hughes and Powell had sixty tons of radioactive sand
brought to a Hollywood studio, where Wayne, Hayward, and
a smaller cast and crew continued filming.

In the decades following the shooting of *The Conqueror*,

there was an unusual series of illnesses and deaths. Susan Hayward developed lung cancer and died in 1975 of a brain tumor. John Wayne also got lung cancer, from which he died in 1979. Cancer killed Dick Powell, as well as supporting actress Agnes Moorehead. Moorehead was later said to have spoken of "radioactive bacteria" that had infected her on the set of the Genghis Khan epic. A character actor named Pedro Armendáriz, best known from the James Bond film *From Russia with Love*, shot himself in the heart while hospitalized for cancer. Moreover, children of Wayne and Hayward wlho had visited their parents in Snow Canyon also developed the deadly disease.

This may be coincidence. No connection has ever been proven between the Harry test and the high incidence of cancer among those connected with the film. Wayne and Hayward were heavy smokers. Moreover, the prevalence of well-known victims often tempts people into conclusions that don't always hold water, as was the case with a series of deaths that came in the 1920s after the spectacular discovery of King Tutankhamun's grave in Egypt. The man who financed the expedition, Lord Carnarvon, suddenly died of a lung infection, and within the space of seven years, seventeen more expedition participants also perished, several succumbing to mysterious infections and five committing suicide. Colonel Aubrey Herbert, Carvarvon's half-brother, died of an abdominal membrane infection, while Lady Almina Carnarvon fell victim to blood poisoning after a mosquito bite. The tabloid press began to speak of "King Tut's Curse," with many authors suggesting that Tutankhamun was reaching out from beyond the grave to punish those who had violated his final resting place. Years later, the American magician and rational skeptic James Randi buried this myth by counting the number of people who survived those seven years. The non-Egyptian members of the expedition had lived on average for twenty-three years after the opening of the grave. Their average age of death, seventy-three, was significantly higher than normal life expectancy at the time. Moreover, the head archaeologist Howard Carter had died peacefully at the age of sixty-six, two decades after his sensational discovery. If there had been a pharaoh's curse, he would surely have been one of its victims.

In the case of *The Conqueror*, *People* magazine determined that ninety-one out of 220 members of the cast and crew had developed cancer. As of 1980, forty-six had died of the disease. *People* magazine may not usually be the most reliable statistical source, but these figures have been accepted by Greenpeace and John Wayne's biographers. If we assume they are accurate, they would suggest a connection between the film shoot in Snow Canyon and cancer. Normally, in a group of 220 people, thirty to forty people at the most would develop the disease. It would be interesting to know how many of the thousands of Native American extras developed cancer, since the battle scenes would have exposed them in particular to radioactive fallout, but no one has studied this question.

There is absolutely no doubt, however, that Howard Hughes personally knew many of the cancer victims from the cast and crew, including the film's director and its two leads. Their fates must have made the billionaire acutely aware of the potential consequences of the atomic tests in Nevada. Hughes may have earned a lot of money from America's nuclear program, but he found nuclear tests very unsettling. He feared that they would scare off tourists from making the trip to his entertainment palaces in Las Vegas. Moreover, as a resident of the city, Hughes felt himself to be at risk. An apocalyptic warning that he issued ahead of the "Boxcar" atomic test showed just how fearful the increasingly paranoid billionaire had become:

> If the gigantic nuclear explosion is detonated, then in the fraction of a second following the pressing of that fateful button, thousands and thousands, and hundreds of thousands of cubic yards of good potentially fertile Nevada soil and underlying water and minerals and other substances are forever poisoned beyond the most ghastly nightmare. A gigantic abyss too horrible to imagine filled with poisonous gases and debris will have been created just beneath the surface in terrain that may one day be the site of a city like Las Vegas.

Such fears spurred Hughes on to action. At first he delayed the construction of one of his weapons factories. Later he tackled the problem head-on. "Alerted to the dangers of nuclear testing by radiation sickness suffered by members of his production crew filming the movie *Conqueror* in Nevada," writes historian and ecologist Dean W. Kohlhoff, "Hughes launched an all-out crusade to stop subsequent tests there." The billionaire had ecologist Barry Commoner of the University of Washington flown in to Las Vegas, where Commoner announced that tests could damage the nearby Hoover Dam. Hughes also personally lobbied major politicians, including presidents Lyndon B. Johnson and Richard Nixon, Vice President Hubert Humphrey, then Governor of California Ronald Reagan, senators Robert and Edward Kennedy, and Nevada Governor Paul Laxalt. But even as the political debate he had unleashed was still going on, Hughes hit upon what seemed to him to be a far simpler solution to the problem: Alaska. Nuclear tests, the billionaire decided, did not have to be forbidden. They only had to be moved to somewhere more suitable—and suitably far from Nevada. According to Kohlhoff, Hughes even offered to pick up the costs for the change of venue. In 1968, a large-scale underground test nicknamed "Milrow" was, in fact, carried out in Alaska. Hughes's role here is debatable. Despite the billionaire's direct intervention with the American president, the Boxcar test took place as scheduled in Nevada. Subsequently, though, Hughes's wishes began to coincide with the preferences of many high-ranking military officers and the Atomic Energy Commission (AEC) for staging future tests in the Great White North. Hughes could not order a test to be moved, but as one of America's leading arms manufacturers, he had excellent connections. He even succeeded in winning over an Alaskan candidate for the U.S. Senate, Mike Gravel, to his cause. On a Hughes-owned television station, Gravel lobbied for Alaska to host a nuclear test. After getting elected in 1969, Gravel quickly reversed his position and became a prominent opponent of nuclear testing in Alaska—a mark of how unpopular Milrow was in America's northern wilderness. The ecological concerns that Hughes had highlighted with such outrage in Nevada were all the more grave in Alaska.

Milrow wasn't the first test to be carried out in America's northernmost state. Attempts to use Alaska as a venue for nuclear experiments had begun in the 1950s. At the time, the U.S. military was looking for a suitable site for America's first underground test. The wilderness in the interior of the state was proposed, but it was quickly rejected as being too dangerous for Native Americans and trappers. Instead planners focused on a remote Aleutian island called Amchitka. This thin, sixty-eight-by-six-kilometer stretch of land possessed an airstrip and a military base from World War II. That made the island attractive to planners, as did that fact that, since the war, it had been uninhabited. Amchitka, however, hadn't always been deserted. A variety of peoples had lived on the island over the centuries, including Aleuts as many as 4,500 years ago. Well into the modern age, Amchitka's natural resources and harsh beauty had attracted people. Russian fur traders had arrived on the island in the eighteenth century and done well for themselves. There was little chance of starving, despite the remote location. The island had excellent fishing grounds just off its coast, and the shoreline teemed with sea otters. Indeed, the combination of rough-hewn hills and lakes made Amchitka one of the most impressive sights in the Aleutians. In 1913, President Taft had made it the core of the "Aleutian Island Reservation," one of the United States' first nature reserves.

Nonetheless, despite the island's protected status, the AEC concluded that it was a "logical" site for a nuclear test. In addition to the factors just discussed, planners found that since Amchitka was relatively small, a limited amount of territory would be contaminated. In a U.S. Department of Energy memo from 1950, it was argued that remote sites like Amchitka and the Eniwetok Atoll in the South Pacific were especially suitable "for tests where the radiological hazards involved may be beyond the limits acceptable in the United States."

In November 1950, President Truman bought these arguments and authorized two tests, one above and one below ground. "Operation Windstorm" was designed to test a nuclear warhead that would burrow beneath the earth's surface before exploding. As so often was the case during the Cold War, the U.S. was very concerned about falling behind the Soviets technologically. Despite the momentum that began to build as a camp of 500 people was set up to plan to test, some high-ranking military officers began to question the value of the project. One admiral deemed Amchitka's weather too poor and its tectonic instability unsuitable for taking seismic measurements. The skepticism went all the way up to the Department of Defense. One DOD expert wrote that an underground detonation, which had never been tried before, carried incalculable risks. It was feared that the explosion might trigger an artificial earthquake. The idea was not completely far-fetched. The Soviets later pursued research aimed at developing just such a seismic weapon. The Kremlin hoped to build a bomb that would be capable of unleashing an earthquake along the San Andreas Fault in California and destroying cities like San Francisco. That idea was eventually abandoned as impractical.

The Windstorm tests were cancelled in July 1951, but this was not due to environmental concerns, and the area remained under threat. Meanwhile, other tests were revealing how problematic the experiments foreseen for the island could be. In November 1951, the U.S. military detonated an underground atomic bomb in Nevada. The "Sugar" explosion had not unleashed an earthquake, but it had gouged a gigantic crater in the desert and thrown up radioactive soil into the atmosphere. The bombs that were supposed to have been detonated on Amchitka were four times as strong as the Sugar warhead.

Alaska was spared its nuclear debut until the early 1960s, but the U.S. military continued to look for a site at which to test whether Minuteman intercontinental ballistic missiles could be destroyed or diverted by the electromagnetic pulse of an atom bomb. Planners needed somewhere above ground where they could detonate a device of about one megaton, and the field of candidates was gradually narrowed to Amchitka and the Brooks Range mountains on the Alaskan mainland. Several members of the search committee preferred the mainland, although it was located near various areas of human habitation, including Point Hope, Noatak, and the town of Barrow. Edward Teller, the "father of the hydrogen bomb," was also lobbying to be allowed to carry out nuclear tests of his own in the region. He'd run into strong local opposition—a fact not lost on the Defense Department.

The focus once again shifted to Amchitka, but once again there was a decisive turnabout, this time on the part of the government and not the military leadership. Washington ordered a moratorium on nuclear testing. The nature reserve on the edge of the Arctic Circle had once more been spared its nuclear fate.

It was a top-secret project, so confidential that not even the AEC was in the know, that succeeded where Operation Windstorm and the Minuteman tests had failed. In 1958, the United States agreed with the Soviet Union to impose the first bans on nuclear testing. However, no sooner were the moratoriums in place than Washington began to worry that the USSR was secretly violating them. By 1963, fears of this kind were so rampant that Pentagon strategists came up with a billion-dollar program called "Project Vela" to detect whether nuclear tests were being carried out. The most spectacular element of the project was "Vela Hotel," a ring of top-secret military satellites that scanned the earth for suspicious emissions of radiation. It attracted worldwide attention on September 22, 1979, when one of the satellites reported a double explosion over the South Pacific. Rumors persist that Israel and South Africa carried out a joint nuclear test on an ocean platform near the Prince Edward Islands. Recently released documents do show that Israel sent nuclear technology to South Africa.

But another part of the project was "Vela Uniform," an initiative aimed at using seismic measurements to reveal subterranean nuclear tests. The problem, however, was that underground shock waves caused by an atomic detonation were indistinguishable from those caused by an earthquake. According to scientific estimates, there were at least 100 earthquakes every year within the Soviet Union that had the same strength as a twenty-kiloton atomic bomb. The key to telling the difference between the two types of phenomena was to discover a seismic "fingerprint" only left behind by nuclear explosions. To do that, scientists needed data, and to get data, they had to carry out an experiment.

Since Amchitka was located in an area of tectonic instability, it was an ideal site for comparing seismic data, and the island quickly attracted the attention of "Uniform" planners. The Aleutian islands sit directly above the so-called Pacific ring of fire, the volcanically active area just north of the Pacific Plate. There, the earth's crust is only a couple of kilometers thick. In 1964, Alaska was rocked by a magnitude-9.2 earthquake—the largest ever recorded in North America—and on February 4, 1953, a magnitude-8.7 quake hit the Rat Islands, of which Amchitka is a part. The unusual strength of this seismic activity worried military planners. Human knowledge of the earth's tectonic plates was still in its infancy in the 1960s. No one could rule out the possibility that a nuclear explosion might trigger further earthquakes in the seismically volatile region. Moreover, it was thought that the hollow subterranean chamber caused by the test, which was code-named "Long Shot," would be subject to tears whenever the earth's tectonic plates moved. That would have allowed radioactive material to leak out, causing certain environmental catastrophe. Nonetheless, despite these fears, Long Shot went ahead on October 29, 1965. The blast created small subterranean fissures through which radioactive isotopes like tritium and iodine 131 leaked out, but contrary to most expert predictions, the damage was limited. Nonetheless, the many mud geysers and landslides unleashed by the detonation hinted at the destruction a larger explosion could cause to this sensitive biosphere.

With a force of only eighteen kilotons, Long Shot was dwarfed by the two later detonations for which it paved the way. Four years later, on October 2, 1969, the U.S. military carried out a massive atomic test on Amchitka. Originally, the exercise was code-named "Ganja," but the AEC soon realized that this was slang for marijuana, and the bomb was rechristened "Milrow." With a yield of one megaton, the exercise was meant to test a new warhead with the capacity to destroy enemy ICBMs in mid-flight. The test was originally supposed to have been held in Nevada, but there the AEC encountered resistance in the form of a political protest movement allied with Howard Hughes's private anti-testing campaign. Moreover, a similar test had left behind a huge fissure on the desert surface, suggesting that Nevada's geology was unsuitable for underground experiments of this magnitude.

Ahead of Milrow, the U.S. military expanded its facilities and roads and built a large drilling tower on Amchitka. The explosion caused by the hydrogen bomb left the island permanently scarred. At Ground Zero, the island's surface was raised some four and a half meters, while ugly fissures appeared elsewhere. Two lakes immediately dried up, leaving fish with shredded air bladders flopping around helplessly on land. Hours after the explosion, small-scale earthquakes still shook the island.

Despite the havoc that Milrow had wreaked on the nature conservation area, the AEC returned to Amchitka in 1971 to carry out the largest underground test in American history. "Cannikin" had a yield of five megatons, the equivalent of five million tons of TNT. Environmentalists were so outraged that they raised several thousand dollars for an old cutter with which they wanted to sail to the test site in an effort to prevent the detonation. The attempt to disrupt the test failed, however, because the activists did not know that it had been postponed. When they arrived at Amchitka, nothing happened, and several days later their cutter was boarded and impounded by the Coast Guard. Nevertheless, crew members were celebrated as heroes, and the name of their vessel, *Greenpeace*, eventually became the title of the world's most famous environmental organization. The Amchitka mission is rightly considered the moment at which the movement was born.

The Cannikin detonation formed an artificial crater two kilometres wide and twelve meters deep, which quickly filled with water. Today, it remains both the largest and deepest lake on the island. On the shore, a large cliff containing an archaeological site with prehistoric artifacts broke off from the rest of the island and tumbled into the ocean. Measurements made at Greenpeace's behest in 1996 still registered significantly elevated radiation levels on Amchitka. Samples of fresh water from the island contained plutonium 239, plutonium 240, and americium 241. It is unclear, however, whether the radioactive isotopes came from the subterranean tests or the many aboveground atomic exercises carried out by the Soviet Union.

In 2004, oceanographers from the University of Alaska returned to the Long Shot and Cannikin test sites and examined the coastlines for traces of radiation. The results of their examination were reassuring. There were no fissures under the water's surface, nor were there deviations in the ocean water's salt content, which would have indicated that subterranean fresh water was seeping through to the earth's surface. The seal provided by the island had remained intact.

Nature, it seems, was able to absorb these three nuclear blows. Nonetheless, the island remains scarred by its military past. Once North America's wildest natural paradise, Amchitka is now mentioned in the same breath as Semipalatinsk, Eniwetok, and the Bikini Atoll.

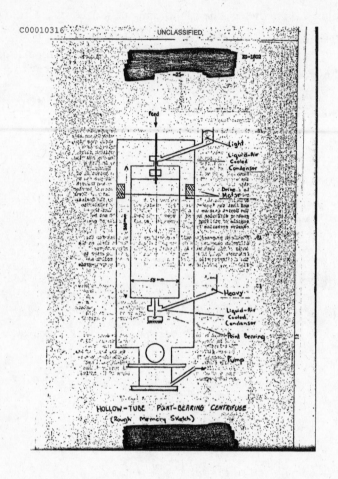

HOLLOW-TUBE POINT-BEARING CENTRIFUGE
(Rough Memory Sketch)

A sketch of a Zippe centrifuge from a top-secret CIA file. This device was instrumental in the spread of nuclear technology throughout the world.

The Soviets tested dirty bombs on this German torpedo boat, renamed the *Kit*. In the process they contaminated Europe's largest lake.

Today, clumps of weapons-grade plutonium dot the landscape of the Polygon test site in Kazakhstan. No one knows whether any of the potentially lethal material has fallen into the wrong hands.

Operation Crossroads aimed at sinking an entire naval fleet. The "Baker" test caused a 600-meter-thick column of water to shoot up into the air. The test destroyed coral on the ocean floor and contaminated the Bikini Atoll.

During the test "Dog," soldiers from the 11th Airborne Division watch the mushroom cloud from a distance of about eleven kilometers.

The shock wave from the test "Stokes" crushed the frame of an unmanned ZSG 3 airship.

The U.S. Army's Davy Crockett rocket launcher fired a nuclear warhead that weighed only thirty-four kilos.

John Wayne on location as Genghis Khan in radioactively
contaminated Snow Canyon, Utah.

CHAPTER FIVE
Swords into Plowshares

The Cold War was an era in which everything had to be made higher, faster, and bigger, regardless of the cost. On both sides of the Iron Curtain, overheated masculine minds worked feverishly on projects that would be sold as leaps forward in human history. The gods of war were whispering in scientists' ears, and they came up with things previously beyond the human imagination. Astronauts cruised the surface of the moon in lunar jeeps. Rocket-propelled vehicles shattered all previous speed records. Child Olympic athletes were pumped so full of steroids that they grew beards. The global muscle-flexing knew neither natural nor aesthetic limits.

One of the most cockamamie ideas of the period came from a group of American nuclear physicists. In the wake of the Suez Crisis of 1956, the Cold War had spread to one of the world's most volatile regions, the Middle East. Egyptian president Gamal Abdel Nasser began distancing himself from the West and advocating a united, independent pan-Arab state. He was supported by the Soviet Union, which had signed a number of lucrative arms deals with Egypt. Meanwhile, Nasser also recognized the Communist People's Republic of China at a time when the conflict over Taiwan was threatening to boil over. Angered, the United States and Great Britain withdrew an offer to finance the building of the Aswan Dam, a

hugely prestigious project in Egypt. Nasser reacted by nation-
alizing the Suez Canal, a vital shipping lifeline that connected
the Mediterranean Sea with the Red Sea. Israel, Britain, and
France attacked Egypt. Military hostilities were over in a matter of days, but the canal remained impassable for commercial
shipping for much longer. It was a catastrophe for global trade.

In the shadow of events in the Middle East, a small group
of scientists convened at the Lawrence Livermore Laboratory
in Livermore, California, to discuss how similar wars could
be avoided in the future. The solution they came up with was
to build new canals instantaneously by using hundreds of
hydrogen bombs. The logic was this: The United States also
relied on a crucial, manmade waterway on its doorstep, the
Panama Canal. If that passage were to be blocked for political
reasons, shipping traffic from the east coast to the west coast
of the United States would have to be diverted around South
America. That scenario was something Washington was very
keen to avoid. Hence in September 1964, almost a decade after the meeting of the Livermore nuclear scientists, Congress
appropriated 17.5 million dollars for a study on the feasibility of constructing a seventy-three-kilometer canal in Southern Panama by means of "nuclear earthmoving." Researchers
calculated that 302 atomic bombs, lined up in a row, would
be needed for the project. The scientists also suggested some
alternative locations. The United States could blast the canal through another friendly country—for instance, Mexico.
Breaching Mexico's Tehuantepec Isthmus would require 875
hydrogen bombs, a route through Nicaragua, 925. The costs
for each of the hypothetical canals were $665 million in Pan-
ama, $1.85 billion in Nicaragua, and $2.275 billion in Mexico.
The study concluded that, in terms of expense, nuclear earthmoving was clearly preferable to "conventionally dug" canals.

It was down to one man that proposals like these were not only taken seriously, but came amazingly close to becoming reality: the father of the hydrogen bomb, Edward Teller. Not only was Teller present at the original meeting of physicists in Livermore; in the years that followed, he tirelessly lobbied for "Operation Plowshare," the United States' "peaceful" nuclear program. In addition to the Central America canal project, Teller promoted a host of other ambitious ideas to use atomic weapons for things like searching for oil or rechanneling major rivers. He was also fascinated by the notion of producing water on the moon. If only 1 percent of the rock on the earth's satellite consisted of water, a proposition Teller found plausible, "It might be possible to take a nuclear explosive to the moon, detonate it underground, and collect the water produced by the explosion for use by astronauts on a lunar base," wrote Teller and others in a book entitled *The Constructive Uses of Nuclear Explosives*. "As the cost of transporting material to the moon will be exorbitant, this would be the least expensive way to provide water. For obvious reasons, we might dub this Project Moses." And Teller was not alone in his fantasies about detonating a nuclear bomb on the moon: both the United States and, less seriously, the Soviets had considered projects to bomb the moon.

Operation Moses reflected the extreme mechanistic world-view of the man who thought it up. Yet it would be a mistake to view Teller simply as a megalomaniacal lunatic. He was a complex man whose personality had been shaped by a variety of drastic changes in his life. Teller was born in Hungary in 1908 to the Jewish lawyer Max Teller and the pianist Ilona Deutsch-Teller. As a small child he had difficulty learning to speak, but at the age of only four he was already able to calculate how many seconds there are in a year. The end of

World War I and the rise to power of the Hungarian Communists under Béla Kun were a shock to the mathematically gifted youngster. "The Communist Party included perhaps one-tenth of the Hungarian people," he would later write in his memoirs. "Only the communists' discipline, organization, and disregard for law enabled them to gain control." The change in Hungary's political landscape hit the Tellers hard. Because he was a lawyer, Edward's father was considered a class enemy and prohibited from working. Soldiers were quartered in his office, which the Communist authorities regarded as "useless." Hard currency in Hungary was confiscated and replaced with worthless Communist script.

The events that would cement Edward's political convictions came during World War II. By that point, Teller was living in the United States. He had fled there via Copenhagen and London after being horrified by the Hitler-Stalin Pact of 1939 and by the fact that his friend and fellow physicist Lev Landau, a committed Communist, had been imprisoned as a "capitalist spy" in the Soviet Union. When the Red Army marched into Hungary at the end of World War II, soldiers discovered Teller's gravely ill father Max confined to his bed in Budapest. Miraculously, the old man had survived the Holocaust. Yet as Edward later told it, Red Army soldiers had nothing better to do than to plunder Max Teller's apartment, stealing the pillow from beneath the sick man's head.

The Teller family had personally experienced Stalinism, and Edward Teller viewed the rulers of the Soviet Union with an icy realism foreign to many American intellectuals and "champagne socialists." With ceaseless energy, he lobbied a half-dozen U.S. presidents to equip America with ever newer, ever more destructive super-weapons as a deterrent to the Red peril. He was often mocked as a real-life Dr. Strangelove, the mad scientist portrayed by Peter Sellers in the Stanley Kubrick film of the same name. Hatred of Communism was the driving force in Teller's life, his one true constant.

Teller was extremely ambitious, stubborn, and ruthless when it came to achieving his aims. He was one of the first people to comprehend the enormous destructive potential of nuclear fusion, the same process by which the sun generates light and heat. In 1941, the Italian physicist Enrico Fermi suggested to him that the energy of an atomic bomb could be used to melt light hydrogen atoms and create a fusion weapon. At first, Teller considered the idea impossible, but he soon realized that the technical obstacles could indeed be overcome. Even before the first functional atomic bomb was built, Teller was vigorously lobbying for a core-fusion hydrogen bomb. His wartime colleagues at Los Alamos National Laboratory often found him annoying. Nuclear physicist Robert Serber termed him "a disaster for any organization," while others considered him a scarcely tolerable prima donna. He often quarreled with his direct superior, Hans Bethe. Teller was quite insulted that Bethe, and not he, had been chosen to head the division of theoretical physics at Los Alamos, and he also felt that he was being assigned senseless tasks. When Bethe charged him with making calculations concerning the implosion of plutonium, Teller blew his stack. He thought there were others at Los Alamos better suited to carrying out this task and feared that, if he took on the job, he would not be involved in any meaningful way with the bomb that would be used in World War II. Teller refused the assignment. Bethe insisted. Compromise was impossible. "Although I began explaining all those reasons to Bethe," Teller wrote, "he was convinced that I needed to tackle the job; I was just as convinced that if I did, I would make no contribution to the war effort . . . Although Hans did not criticize me directly, I knew he was angry."

To the astonishment of all their colleagues, the head of

the Manhattan Project, J. Robert Oppenheimer, did not discipline Teller. Instead, the rebellious Hungarian was allowed
to select his own research area and was freed from the oner-
ous tasks handed down by Bethe. While hundreds of physicists, chemists, and technicians were working away in the New
Mexico desert on an atomic-fission weapon, Teller was already
a few steps ahead. He was completely devoted to his "baby,"
the hydrogen bomb.

Year later, on October 31, 1952, Teller finally achieved his
goal, and the hydrogen bomb "Ivy Mike" was detonated on
the Eniwetok Atoll. Teller had proved his point to all of his
colleagues, including Oppenheimer, who had become increasingly skeptical about the hydrogen bomb project after World
War II. Teller avenged himself on his former mentor in 1954
with incriminating testimony before the House Committee
on Un-American Activities. Oppenheimer's security clearance
for military projects was subsequently revoked. The scientific
community was appalled at Teller's act of betrayal.

Teller was never able to win over the hearts of the general
public either. Even his magnum opus, the hydrogen bomb,
was viewed with collective unease. The prospect of World
War III hung over humanity's head like a sword of Damocles.
It took less than a year after the Ivy Mike test for the Soviet
Union to develop a hydrogen bomb of its own. America's
tactical advantage was gone almost as soon as it arrived. For
the general public, Teller seemed to be ratcheting up a nuclear
arms race that sooner or later would end in annihilation.

Thus the father of the hydrogen bomb, a man with a desperate need for social acknowledgment, had good reason to try
to improve his public image with a new, less martial project.
Operation Plowshare, America's peaceful nuclear bombs program, was tailor-made to this end. The name was consciously

chosen to conjure up the Biblical phrase "beating swords into plowshares." Teller hoped that a bomb program in the service of massive construction projects would resolve the central ethical dilemma of his life. Like Oppenheimer and all the other scientists involved in the Manhattan Project, Teller had a role in the deaths of hundreds of thousands of people in Hiroshima and Nagasaki. Worse still was the knowledge that atomic and hydrogen bombs were capable of exterminating the human race. When the Manhattan Project was called to life in 1942, moral questions remained very much in the background. The front lines were clearly drawn. On one side was Hitler, whose ranks of gifted scientists had the skills to build an atomic bomb. On the other side were the world's civilized nations, who could not allow the Third Reich to emerge victorious. Only after Nazi Germany surrendered did people begin to ponder broader moral ramifications, and even after the end of World War II in Europe, scientists in Los Alamos continued to work on nuclear weapons.

Those who favored dropping the bomb on Japan argued that it would shorten the war and save American lives and that it would be a frightening demonstration of power that would deter future wars, especially nuclear ones. The latter, cynical calculation indeed has something to it. The Cold War never developed into a "hot" conflict between nuclear superpowers, and the images of horribly burnt mothers and children in the devastated city of Hiroshima are permanently etched in collective human memory. Such nightmare visions are undoubtedly a major reason why 1945 was the last time an atomic bomb has been used as a weapon.

None of the Los Alamos scientists was to blame for the decision to drop Little Boy and Fat Man. The U.S. military commissioned the bombs, and it was responsible for the order that they be used. That at least was the logic followed by Teller, Oppenheimer, and others. Nonetheless, Teller may well have had creeping doubts about the "innocent scientist" argument. That sort of reasoning would have been far more persuasive had the Los Alamos researchers been working on a "neutral" invention that could be used for both military and peaceful purposes. The bow and arrow were invented for hunting so that human beings could feed themselves. The fact that the bow and arrow could also be used to kill other people was not something for which the inventor could be held responsible. Unfortunately for the nuclear physicists, the situation was not analogous for nuclear weapons. The only purpose served by an atomic bomb was to kill great numbers of people.

Operation Plowshare seemed to offer a way out of this ethical dilemma. In Teller's fondest dreams, the U.S. government would charge him with digging canals and harbors, creating *ex post facto* a peaceful, civilian use for his invention. Teller's visions for the Plowshare program were the expression of this misguided need to justify what he had done. In his conviction that every invention could be put to both peaceful and warlike purposes, Teller cited the case of nylons, an example as equally misplaced as the bow and arrow. "Nylon is a weapon in the battle of the sexes," the scientist proposed in 1974 on the PBS talk show *Day at Night*. "This, most people would say is a peaceful invention. But when DuPont first succeeded in producing nylon, the gals weren't getting any stockings because we were at war, and the nylon, all of it, was used in parachutes. It was used in war. The inventor of nylon had no idea it would be used as a weapon, and when it came, he had no control over it whatsoever." Teller's attitude was that it was the task of science to advance human knowledge and give people more influence over nature, but that it was the job of those in power to decide how to use or misuse scientific discoveries. He also voiced the hope that nuclear energy would never again be "misused."

However, Teller's rhetoric of peace could not obscure the fact that Operation Plowshare was a continuation of the Cold War with different means. The sheer idea of using the hydrogen bomb to create American-controlled access to the Pacific Ocean or the Red Sea made the program a political weapon. The initiative also offered anti-Communist hawks within the government a way of circumventing bans on nuclear testing. Teller was part of these right-wing circles and regarded every test moratorium as a threat to the security of the United States. He also took the naïve and horrifically dangerous concept of a "peaceful" super-bomb completely seriously. He knew, however, that there was no way he would be allowed to begin with a gigantic project like a new Panama Canal. Before he could operate on that scale, he would have to provide a smaller, straightforward example of the civilian uses of a nuclear bomb.

In July 1958, Teller appeared without warning in Alaska and revealed an adventurous plan to a group of hastily assembled local politicians. A manmade harbor was to be created in the extreme north of the state that would allow ships to transport the raw materials in the region to more heavily populated areas. A 610-meter-long and up-to-ninety-meter-deep canal would lead to an expansive port. The massive earth removal required by the project would only take seconds—thanks to the use of five atomic bombs. The bombs' explosive force would be equivalent to 2.4 million tons of TNT. The name of the endeavor was "Project Chariot," and the icy headland of Cape Thompson had been chosen as Ground Zero. The mountains in the area were full of valuable black coal that was just waiting to be mined. The new harbor in the far north, Teller promised, would also be a boon to the fishing industry since it would open up previously unexploited areas of the Pacific Ocean. The costs of constructing the port, estimated at 50 million dollars, would be picked up by the AEC. If people wanted, Teller boasted to the visibly impressed politicians, he "could dig a harbor in the shape of a polar bear."

To many people, Project Chariot looked like a blessing to the Alaskan economy and a solution to Alaska's perennially tight state finances. In the wake of a similar appearance Teller made in Fairbanks, the *Fairbanks Daily News-Miner* rhapsodized that the project was a "fitting overture to the new era which is opening for our state." Chariot was touted as a gift from the federal government, but it was not nearly as cheap as it initially seemed. Two Teller associates who had been sent to Anchorage revealed to a group of local businessmen that the private sector and the state government would be responsible for constructing port facilities, which would cost between 50 and 100 million dollars. Their audience was disconcerted, and Teller was compelled to correct his associates' estimates the following day, reassuring the citizens of Alaska that 50 million was the maximum they would have to come up with.

Teller's reassurance, through, contradicted estimates drawn up by his very own institute, and this wasn't the only occasion on which he would make misleading statements about Chariot. A harbor that far north would have been frozen over nine months out of the year, and there were no significant fish stocks in the waters off Cape Thompson. So the project's benefits for the Alaskan fishing industry were illusory. Even more absurd was the idea of using the port to transport coal. The Brooks Range, a desolate chain of mountain peaks, did in fact contain large amounts of the raw material, but they were located 400 miles away from the proposed site of the harbor. To bring the coal to sea, Alaskans would have had to build a train line under the harshest conditions, and laying tracks in the Arctic permafrost would have cost at least 100 million dollars. Project Chariot thus made no economic sense whatsoever.

Moreover, there was one other minor detail that Teller's group of scientists hadn't gotten right. Cape Thompson had been selected as a location because it was presumably devoid of people. The region, however, was not nearly as uninhabited as the nuclear demolitions experts had claimed. Three hundred and seventeen members of the Inupiat tribe lived only fifty kilometers away from Ground Zero. Their village was called Point Hope. Sixty-four kilometers distant from Ground Zero was the town of Kivalina, with 146 inhabitants, and fifty-six kilometers farther on was the Noatak settlement, where 270 Inupiat lived. Even a detonation with half the planned strength of Chariot would, according to a statement made by the later head of the Plowshare program, have required a quarantine zone of at least 112 kilometers. In response to economist George Rogers, who asked what the Inupiat were supposed to do if they could no longer hunt aquatic mammals and caribou near the harbor, Teller said that they would have to change the way they lived. As soon as the port was finished, new jobs in the Brooks Range would need to be filled. The Inupiat could mine coal.

At this point, none of the Inupiat in Point Hope suspected that plans were underway to turn them into coal miners. Gradually, though, information about Project Chariot began to circulate. While tracking caribou near a river called Ogotoruk Creek, Inupiat hunters encountered one of Teller's vanguard expeditions. The scientists claimed to be examining the geology of the region at the behest of the AEC. They didn't reveal that the traditional Inupiat hunting grounds were being scouted as a potential Ground Zero for a nuclear detonation 100 times as powerful as the bomb dropped on Hiroshima. Rumors nonetheless started to fly fast and furious.

Since no official announcements about Chariot were made outside the main population centers in Southern and Eastern Alaska, it wasn't until 1959 that the Inupiat learned that their worst fears were being confirmed, and the news came neither from the AEC nor Teller's Livermore Institute, but from a traveling missionary from the town of Kotzebue who had read about Teller's plans in the press. After showing the Inupiat a Christian film, the pious traveler revealed that their hunting grounds were going to be blown up and contaminated with radiation. Some of those present had heard about the devastating consequences of American atomic tests in the South Pacific. The Inupiat may have lived in relative isolation, but they were by no means ignorant and even less stupid. So on November 30, 1959, the Point Hope village council sent an outraged letter to the AEC. That communication made it clear that "we ... do not want to see the explosions at near areas of our village Point Hope for any reason and at anytime [sic]."

Meanwhile, the masterminds of Plowshare had other concerns. Since late summer 1958, the United States had completely ceased all nuclear testing in response to a previous unilateral moratorium declared by Khrushchev in the Soviet Union. Despite bitter complaints by Teller and others, the American test ban held for three years. Still, all was not lost for the scientists at Livermore. Their project, as they repeatedly stressed, was for peaceful purposes, and it was debatable whether a moratorium applied to non-military uses of nuclear weapons. Naturally, those in charge—particularly the AEC—knew that a nuclear detonation in Northern Alaska would yield knowledge valuable to the military. That was one reason that the preparations for what was euphemistically called the "shot" continued apace, with the blessing of the higher-ups. Yet since there was no way to plausibly argue for the economic utility of the manmade harbor, and no one stepped up to cover the horrendous costs of building the port facilities and the railway, Teller's men changed their plans. In February 1959, they ceased talking about creating a port and the accompanying infrastructure in the far north and instead said that "Chariot" was merely intended to "demonstrate" what nuclear earthmoving could do. In other words, the AEC was prepared to sanction a nuclear detonation to gain insight into practical questions like how large a nuclear warhead needed to be to create a crater of this or that dimension. Teller's grand, idealistic vision, with which he had tried to attract supporters, suddenly shrank. Instead of detonating hydrogen bombs, several "normal" atomic bombs, with only 19 percent of the originally planned yield, were now deemed sufficient. This was no consolation to the Inupiat. What the scientists didn't mention was that atomic bombs are much dirtier than hydrogen bombs. Nuclear chain reactions release extremely radioactive fission by-products. Compared to their explosive power, the emission of dangerous radiation is much higher in fission bombs than in fusion-based hydrogen bombs. Teller had repeatedly boasted that his project would be "clean," a claim that became utterly absurd when the plans were revised.

Teller typically compared the amount of radiation to which the average Alaskan would be exposed with the dangers generated by wearing a wristwatch with a phosphorus face. But his plans were anything but harmless. Atomic detonations that leave behind craters are much more damaging to the environment than those that take place completely above ground. The detonation of a warhead buried slightly below the earth's surface propels enormous amounts of dirt and rock up into the air, all of it radioactively contaminated. Millions of tons of radioactive material would have been blown into the winds. Even worse, the extreme north of Alaska had an extremely sensitive ecosystem, and radiation would have progressed rapidly up the food chain. Lichens would have sucked it up like sponges. Caribou and seals would have eaten the lichens and then been eaten in turn by the Inupiat. Thus, in no time, lethal elements like strontium and cesium would have entered the human food supply. Moreover, in 1959, those who lived within the Arctic Circle were already subject to radiation levels clearly in excess of the North American norm, since radiation from aboveground tests in the Soviet Union tended to collect at the poles.

Knowledge of how radioactivity could cause cancer and damage human DNA if it got into the food supply was widespread in the late 1950s, so it was no surprise that resistance began to form in Alaska to Teller's plans. The *Fairbanks Daily News-Miner*, the very newspaper that had greeted Chariot with such initial enthusiasm, began to print letters from concerned citizens. Rumors also began to spread within the state's scientific research community. The president of the University of Alaska wrote a letter to the AEC, passing along the reservations expressed by several biologists about Chariot's aftereffects. The organization reacted with a PR campaign and tried to undermine the criticism by co-opting skeptical scientists in the project. The first quarterly report of the AEC in 1959 read:

> A Planning Committee on Environmental Sciences has been established to recommend studies aimed at accomplishing a biological survey of the Cape Thompson and of Alaska. In these studies, information will

be compiled on marine and land population densities, migratory habits, food chains, oceanography, and other pertinent subjects, as part of an investigation to enable the Commission to make the necessary evaluation as to the feasibility of project CHARIOT, a proposed experimental nuclear detonation near Cape Thompson which would provide an excavation suitable for a harbor.

John N. Wolfe, the head of the environmental section within the AEC's division of Biology and Medicine, was put forward as the head of the planning committee.

All told, the AEC handed out $382,098 to forty-one universities, including the University of Washington, the University of California, and the University of Alaska. But if the AEC hoped that the scientists at those institutions could be bribed, they were mistaken. The third quarterly AEC report for 1959 already intimated that Chariot might have to be scuttled:

When Project CHARIOT, Phase II, was approved by the Commission on May 22, 1959, it was also determined that by January 1, 1960, there would be a recommendation relative to whether the project should proceed or be indefinitely suspended. The recommendation is to be based on the results of the summer surveys and other factors applicable at that time.

In the meantime, Wolfe had set up camp with several dozen scientists from various disciplines in an improvised facility on Ogotoruk Creek. The researchers lived in Jamesway huts, circular constructions of waterproof cotton stretched across bent wooden skeletons. The wind howled down from the desolate peaks into the river lowlands. For centuries, Inupiat hunters had used the spot by Ogotoruk Creek as a campsite. Now the AEC had commandeered it. The Inupiat were less than amused by the intruders. There was an angry confrontation in which a resolute woman from Point Hope excoriated the outsiders as "bringers of evil."

Wolfe stuck to his guns, publishing an initial preliminary report on December 10, 1959, that recommended going ahead with the detonation the following spring. At that time of year, he argued, much of the local fauna would be hibernating under a blanket of snow, and there would be hardly any Inupiat hunting. The independent scientists, who had only just commenced their work, were outraged. Wolfe, they felt, was trying to jump-start the process without waiting for the results of their investigations. The biologists A.W. Johnson, Leslie A. Viereck, William O. Pruitt, and L. Gerard Swartz from the University of Alaska denounced Wolfe's report, and a young geographer named Don Foote wrote an angry letter to the AEC, disputing Wolfe's claim that the Inupiat didn't hunt much in the spring. On the contrary, Foote wrote, Inupiat hunters were particularly active during that time of the year. The press, too, began to take an increasingly skeptical view of Chariot.

The AEC saw itself compelled to initiate direct contact with the populace of Cape Thompson and organized an information session by Chariot planners in Point Hope. Luckily, for posterity, the Inupiat made recordings of this event, which were later uncovered by historian Dan O'Neill. On the morning of March 14, 1960, a small Cessna landed in the Arctic Circle. On board were AEC Security Coordinator Charles Weaver, Head of Technology Russell Ball, PR specialist Rodney Southwick, and a zoononic disease expert named Robert

L. Rausch. They were received in the cramped Point Hope community center, in which about 100 Inupiat were sitting silently, their backs to the walls. The atmosphere was tense. In an attempt to relax the mood, the AEC representatives showed a short publicity film about the advantages of peaceful nuclear detonations. To crown the absurdity, the government's PR men had even included a cartoon simulation of the explosion at Ogotoruk Creek. It replaced the gigantic clouds of dust that would have resulted from the detonation of five nuclear bombs with small, friendly-looking clouds.

After the presentation, the AEC asked if the Inupiat had any questions. The local pastor Keith Lawton spoke in detail about the community's concerns vis-à-vis the seismic effects of the detonation and the long half-life of the nuclear waste that would be created. Ball yielded the floor to Rausch. Although as an expert in diseases caused by parasites he was not, in any sense, qualified to give information about questions of nuclear physics, Rausch assured his audience that the fallout would hardly be measurable. There would be no danger whatsoever to either human beings or animals, and all the radiation would have disappeared from the area after a few months. The other AEC experts made no attempt to contradict these statements, although they knew that they were false. Rausch tried to parry critical follow-up questions with further lies, but the listeners' resistance to what they were hearing did not dissipate. Southwick attempted to call an end to the event.

At this point something unexpected happened, and the mood threatened to get entirely ugly. While Southwick was thanking the village leader David Frankson for his cooperation, the latter's wife Dinah interrupted him with a long statement in Inupiaq. Another village resident, Daniel Lisbourne, translated.

Lisbourne: Ah, the woman here, mentioned, all of these people here, all these people, most of them are just silent right now and they have great fear in, in this detonation and the effects, and how the effects of it will be."

Southwick: Internationally?

Lisbourne: No, here.

Ball: What? I don't quite get what her question was?

Tommy Richards (pilot): The effect of the blast.

Ball: On your own Eskimo people? Oh, well, ah. I believe we've covered that already ... [Ball turns to Richards to discuss the return flight.]

Richards: I don't think that's true. You haven't.

Several people then began talking excitedly in Inupiaq. The AEC men spent the next few hours dealing with a series of pointed questions. It wasn't until 5:00 p.m. that the crowd allowed them to leave Point Hope. Before their airplane had disappeared from view in the sky, the village council took a vote on Chariot. It was unanimously rejected. One year later, several of the scientists involved the project, including Viereck and Foote, publicly excoriated the mendacious and irresponsible behavior of the AEC.

On April 30, 1962, as a result of pressure from the media, scientists, and the Point Hope village council, Project Chariot was cancelled. The initiative had been the centerpiece of

Operation Plowshare, yet in a book about it that Teller would later write together with three other main directors, he and his co-authors devoted only a single page to their plans for using nuclear weapons to create a manmade harbor in the Arctic Circle.

Sadly, that was not the end of the story for the people of Point Hope. In 1992 it emerged that AEC scientists had buried the radioactive substances iodine 131, strontium 90, and cesium 137 at various locations on Ogotoruk Creek and then poured water over them in order to measure the extent to which nuclear fallout would spread in the water supply. The scientists had no official permission to experiment with these substances, and the quantities set free were well in excess of accepted limits at the time. The Point Hope village council was outraged. On October 17, 1992, the following press release hit the newswires:

ATOMIC ENERGY COMMISSION OFFENSES AGAINST THE PEACE AND SECURITY OF THE INUPIAT OF POINT HOPE

We, the Inupiat of Point Hope, have the ability to face the arrogant policies of the former Atomic Energy Commission and its Project Chariot. We will not be willing victims for the genocidal and inhuman policies of the Nuclear Energy Commission. Our Grandfathers and Grandmothers have taught us to persevere through hostile circumstances and unforeseen environments. Our ancestors had informed us of our spiritual dimensions and taught us that our spirituality and strength lies within ourselves. That has been the source of our honesty and respect for others which the Europeans have historically misunderstood as "childishness." In our recent history with the United States, we had not expected such deceptiveness and complete disregard for ourselves as we still held to our traditional teachings. Nor had we expected their hidden policy of involuntary destruction of a People, ourselves, the Inupiat.

On October 9, 1992 one-hundred angry and fearful residents of Point Hope met at their community center. The finality of the community spirit was purely and clearly stated by veteran George Kingik. He said that, "The United States soldiers were used as research subjects to study radiation effects. This is commonly known and it isn't so far fetched to assume Arctic Eskimos were used in the same way. World War II was a holocaust for the Jews but the Jews were treated better than we were. The Jews knew it was coming." For the Inupiat, the destructive force was completely unforeseen and invisible. There was not even the appearance of civilized gesture and involvement.

... The Atomic Energy Commission made a critical shift from demobilization to a pathological experimentation of the radioactive contamination upon ourselves and our pristine homeland. The Atomic Energy Commission failed their public trust responsibility and deceptively left the lethal strontium 90 in the groundwater of Inupiat land for monitored absorption by our beautiful lands and waters and ultimately by ourselves and our teachers, the Animals.

The press release also pointed out that the United States had signed the 1948 Convention on the Prevention and Punishment of the Crime of Genocide and was thus legally bound to follow its provisions.

Shortly after this statement, the radioactive material in Point Hope was located, dug up together with several tons of surrounding soil, and taken to Nevada, where it was sealed underground in that state's nuclear test site. A representative of the Alaskan government, Dennis Roper, reported to the Inupiat that the matter had been settled once and for all.

It's horrific to think what would have happened to the people and the environment of Cape Thompson had the AEC succeeded in pushing through Teller's project. We can get a rough idea of the ecological devastation from some of the other atomic tests carried out as part of Operation Plowshare. With the expiration of the moratorium on nuclear testing in 1961, the "peaceful" atomic landscapers of Livermore were back in business, detonating a series of bombs in the southern United States.

Plowshare kicked off with the relative small "Gnome" test near Carlsbad, New Mexico, on December 10, 1961. It was aimed, among other things, at investigating whether a nuclear explosion could be harnessed to produce energy. But the detonation destroyed the machinery that was supposed to convert the blast into power. Moreover, a radioactive cloud drifted over Route 128, a major north-south highway that followed the Colorado River. The road, of course, had to be temporarily closed.

Far more unsettling were the effects of the "Sedan" shot carried out in the Yucca Flat section of the Nevada Test Site on July 6, 1962, two months after Chariot was officially abandoned. In retrospect, the 100-kiloton detonation seems like a monstrous and utterly irresponsible act of compensation by Teller and his associates for their failure in Alaska. The test was aimed at investigating the use of nuclear bombs for mining and other earthmoving purposes. Although the bomb used was at the time the largest ever to be detonated on U.S. territory, Plowshare directors promoted it as a "clean" explosion, in which contaminated soil would simply fall back into the crater made by the blast. But the AEC's calculations could hardly have been farther off the mark. Sedan created enormous clouds of dust that meandered through various American states in the days after the explosion. Unusually high levels of radiation were measured in the suburbs of Salt Lake City, and carcinogenic iodine 131 was found in milk and baby food. The hardest-hit area was the small city of Ely, Nevada. A sandstorm caused by Sedan left the town so darkened that the street lights had to be turned on in the middle of the afternoon. "As predicted," claimed the AEC.

Sedan's explosion left behind a 365-meter-wide and almost 100-meter-deep hole in the desert floor. The crater was large enough to be visible from the moon. Sedan resulted in the highest levels of nuclear fallout of any American nuclear test, a fact that should have been enough to convince even the staunchest proponents of "peaceful" atomic weapons that there was no such thing as a harmless nuclear explosion. But the dictates of logic were suspended during the Cold War. As a 1983 Department of Energy dossier revealed, representatives of the Department of Defense were always present at the "peaceful" experiments. Military experts had every reason to closely follow the tests. Over the course of the Cold War, the Pentagon developed countless Atomic Demolition Munitions, or ADMs. These were the defensive weapons intended for detonation in crucial passes or logistically key locations to halt the advances of the invading Red Army. It doesn't take much imagination to conceive of how the Plowshare data might have helped the development of the roughly 300 ADMs that were deployed in Western Europe. Tests carried out by the "peaceful" bombers provided crucial information about how to shift massive amounts of earth and create landslides and craters.

Teller, who enjoyed both military and civilian backing, wasn't the sort of man to be discouraged by setbacks. On the contrary, in the wake of Chariot and Sedan, he continued to develop ever more cavalier ideas for "peaceful" nuclear detonations. In 1963, the father of the hydrogen bomb suggested blowing up an entire mountain range in Southern California to create space for a highway and a railway line. Business partners were quickly enlisted, and the local press in San Bernadino, where the most important newspapers belonged to a member of the California Highway Commission, celebrated Teller's plans. The new project was named "Carryall," and it allowed Teller once again to assume the role of the visionary. The AEC, as Teller later recalled, formed a working group together with the California highway authority and the Santa Fe Railway. The aim was to shorten the stretch of railroad between Golls and Ash Hill to 101 kilometers, making it straighter and less subject to rises and falls. The working group suggested moving the tracks a bit to the north and the highway slightly to the south so that both would run parallel through a nuclear-excavated mountain pass. "In many cases terrain need not be the dominating factor in determining highway routes, as will be seen in the case of Carryall," Teller boasted. "There are situations in which it is more economical to design a highway through terrain requiring deep nuclear excavations than through terrain requiring shallower conventional excavations.

Hence routes normally considered economically prohibitive could be undertaken."

Put simply, Carryall aimed at blasting a 120-meter-deep chasm through the Bristol Mountains. To excavate massive mountain rock over a distance of 3.2 kilometers, Teller's Livermore Institute calculated that engineers would need twenty-three bombs with a total yield eighteen times greater than that of the Sedan test. The scientists dismissed concerns about the enormous amounts of radiation released by the explosions. Four days after the detonations, the Institute claimed, workers would be able to enter the site and start constructing the highway. Initially they would require protective clothing. But after a year at the latest, the project could dispense with such safety precautions.

It's not surprising that these prognoses were overly optimistic. But how could the scientists arrive at such crass miscalculations? Teller and his colleagues presumed that researchers would make swift progress in developing "clean" nuclear bombs that hardly released any radiation. That turned out to be wishful thinking. The problem of radioactivity was never overcome—a fact that, however, was not allowed to disrupt the planning of Carryall. Some 30,000 people lived in the vicinity of the Bristol Mountains, where Teller wanted to stage his nuclear firestorm. All of them would have been directly affected by the fallout. An independent study concluded that the radiation from Carryall would have been five times as great and would have spread twice as far as the Plowshare scientists predicted. Yet despite such gloomy assessments, Teller was given the go-ahead to demolish the Bristol Mountains. The project was scheduled for 1967.

Once again, though, Teller's plans were foiled. This time what got in the way was not a local citizens' initiative, but a new nuclear testing moratorium agreed on by the United States and the Soviet Union. The AEC was forced to delay the project by eighteen months, and somewhere in that period, California highway authorities got cold feet. Carryall was never officially cancelled, but by the end of the 1960s it was no longer officially discussed. Still, those who had been involved with the project steadfastly denied that concerns about radiation had scuppered the nuclear highway.

With Teller having failed to demonstrate the practicability of the "peaceful bomb" with even a single concrete example, it didn't look good for his prestige project—the construction of a second Panama Canal. Almost a decade into Operation Plowshare, no one at the AEC could plausibly explain how the Panamanian population and environment could be protected against the devastating effect of radiation. No workable plans had been drawn up to deal with the 40,000 people who lived in the immediate vicinity of the canal's envisioned location. And even if they could have been resettled, there was still the problem of the Cunas, an aboriginal people who lived in isolation in the Panamanian jungle. It was virtually impossible to contact the Cunas because they basically did not interact with the outside world.

Slowly but surely, Plowshare degenerated into a series of senseless tests that only illustrated that atomic bombs were useless for anything but killing and destruction. Teller and his colleagues can't be accused of a lack of imagination, though. In New Mexico, they tested whether nuclear detonation could be used to drill for gas, but the gas that was released during the experiment was far too radioactive to be sold. Other comparably ambitious plans never got beyond the drawing board. They included using nuclear bombs to melt the ice from polar ports, to re-channel rivers or to desalinate salt water from the ocean.

Operation Plowshare carried out twenty-seven tests before it was finally discontinued in 1974. Some of its radioactive by-products will be with us for a very long time. Strontium 90, which is highly carcinogenic, has a half-life of twenty-five years. The half-life of caesium 137 is thirty-three years, and plutonium 239 takes no less than 24,110 years to lose half of its radioactivity.

Teller and his colleagues were not the only ones who engaged in "peaceful" nuclear detonations. The Soviets devoted even more time and energy to a Plowshare-style program of their own. Between 1965 and 1989, they carried out 116 civilian explosions. Thirty-nine bombs were detonated for seismological experiments. Thirty were aimed at creating subterranean storage caverns. Twenty-one sought to uncover oil and gas reserves. Five were used to combat fires at oil fields. And twenty-one bombs were detonated in attempts to create canals and water reservoirs. The detonations took place primarily in Kazakhstan, then a Soviet republic, and Russia itself. The Soviet Union was determined not to fall behind the United States in any area of technology—and especially not in one that Moscow considered so vital. In an uncompromising duel to see who could create the biggest bang, the Communist leadership ordered a Soviet version of the Sedan test. Under the oversight of the head of the Medium Machine Building Ministry and the man responsible for the Soviet Union's entire nuclear program, Efrim P. Slavskiy, a detonation was carried out in the Polygon test site near Sarzhal. It left behind a 100-meter-deep crater that then filled with water. A Soviet propaganda film later showed a fearless swimmer doing laps in the artificial atomic lake. The film commentator explains that "water from the artificial lakes created by nuclear explosions is absolutely safe for human health." No one knows what happened to the swimmer, or even who he was. In the 1990s, a high-level official in the Soviet Foreign Ministry claimed that the swimmer was none other than Slavskiy himself.

Kazakh herdsmen still live in Sarzhal. In 2009, one of them told journalists Yermek Boltayev and Regina Kozhikova that Korean scientists had measured the levels of radiation in the lake and then expressed astonishment that any human beings had been able to survive in the area. The water in the lake contains high levels of tritium, but a study commissioned by NATO concluded that it had apparently not seeped into the wells in Sarzhal, where tritium levels were astonishingly low. That is not to say, however, that the inhabitants of Sarzhal have been spared the consequences of the region's past. Several of the herdsmen told of abnormal numbers of children being born with genetic diseases in the decades following the explosion, though the connection is circumstantial, especially as there are other severe environmental problems in the region.

In contrast to Teller and the "peaceful" bombers at the AEC, Communist leaders could carry out experimental nuclear detonations without regard for collateral human damage. Scientists in the West could never afford to be equally uncompromising—the anti-nuclear movement in the 1960s constantly held them accountable for their actions. The anti-Chariot protests in Alaska were one of the first triumphs of this new social force. Ironically, Teller's project encouraged a paradigm shift in society's relationship to nature and the state. People no longer saw a healthy environment as just a source of excess value. They were also willing to defend it against state institutions. It is hardly surprising that Teller himself reacted with anger to the persistence of environmental activists. "The Russians, who are not as limited as we are by extremists parading under the banner of environmentalists, have moved ahead of us in Plowshare," he complained.

Others around the world also tried to think flexibly and find a peaceful use for atomic bomb—with a similar lack of success. Around the same time as the final years of Operation Plowshare, a German engineer conceived of a gigantic energy project that included a "peaceful" nuclear detonation. Professor Friedrich Bassler was the director of the Institute for Water Engineering at the Technical University of Darmstadt. In World War II, he had served under Rommel in Northern Africa and had gotten to know and love the Middle East. In the early 1970s, he developed a bold plan to connect the Qattara Depression in Egypt by canal with the Mediterranean, eighty kilometers to the north. Since the Depression is located up to 130 meters below sea level, water would have flooded into inland Egypt and created a lake of about 3,500 square kilometers.

Bassler wanted to set up power-generating turbines at the point where the canal fed the lake. The potential energy yield was considerable. The problem was a small mountain range located on the route between Qattara and the Mediterranean. So Bassler suggested using 200 subterranean hydrogen bombs to excavate the canal. The bombs were to be between 100 and 1,000 kilotons in yield, and since Qattara was sparsely inhabited, "only" 25,000 migrant Bedouins would need to be evacuated. The costs for the project, Bassler emphasized, would be moderate, some 370 million deutschmarks (ca. $230 million) for the nuclear material and around 420 million ($260 million) for "carrying out the detonations and the conventional aspects of canal construction on the coast." The total expense of 790 million deutschmarks, Bassler argued, was only around a fifth of that associated with conventional canal construction.

Summing up the positives of his plan, Bassler wrote: "Leaving aside the economic and energy advantages of building the canal, this method would save somewhere along the lines of three billion deutschmarks."

Bassler contacted the AEC about the possibility of procuring the hydrogen bombs he envisioned for his project. The request from West Germany would have been music to the ears of the Plowshare leadership. The project almost went ahead. The will, in any case, was there. Bassler succeeded in attracting supporters both in Egypt and West Germany.

The Qattara project came under heavy fire from the start. The general public in West Germany in the 1970s, which was about to see the rise of the Green Party, was all too aware of the potentially fatal ecological consequences and health risks associated with radiation. From a geopolitical power perspective, it would also have been difficult for the United States to allow Egypt and West Germany, neither of which possessed the bomb, to use nuclear weapons. Bassler was undoubtedly not pursuing any sort of hidden agenda. His plan was a purely civilian one aimed at producing energy and had no military involvement. Nonetheless countries like France and Israel would have definitely felt a trifle uneasy at the thought of a joint German-Egyptian Plowshare project. Such basic flaws doomed Bassler's idea. It died before it had the chance to provoke outrage among the general public in West Germany or abroad. By the end of the 1970s, the Qattara project lost all the support it had once enjoyed in Egypt and disappeared.

CHAPTER SIX
The Doomsday Machine

On April 7, 1954, the *New York Times* ran an article detailing
an apocalyptic scenario. One day a ship would appear off the
American coast, at a distance great enough that it would go
undetected by the U.S. Coast Guard. After receiving a signal
from Moscow, the crew would detonate a new kind of nuclear
weapon. No one in Los Angeles or Washington would notice
that anything was wrong, but after a few days, there would
be mass deaths caused by a gigantic radioactive cloud. The
Times cited an expert in nuclear chemistry, Professor Harri-
son Brown of the California Institute of Technology, who described the fallout from this sort of bomb, if it were detonated
over the Pacific:

> The radioactive dust would reach California in about
> a day, and New York in about four or five days, kill-
> ing most life as it traverses the continent... Similarly
> the Western powers could explode hydrogen-cobalt
> bombs on a north-south line about the longitude of
> Prague that would destroy all life within a strip 1,500
> miles wide, extending from Leningrad to Odessa, and
> 3000 miles deep, from Prague to the Ural Mountains.
> Such an attack would produce a "scorched earth" un-
> precedented in history.

One paragraph further on, the eminent nuclear physicist Leo Szilard calculated that 400 such bombs would be all that was needed to completely exterminate all life on planet Earth. The elements containing the deadly radioactivity would have a half-life of five years—more than enough time to kill even those people who had sought refuge in anti-atomic bunkers.

Szilard knew what he was talking about. He was one of a legendary quartet of Hungarian nuclear researchers whose abilities were so unusual that they were nicknamed "the Martians." All four had very similar biographies. All of them came from Budapest and were born to well-off Jewish families with German cultural roots. All were forced to flee Hungary when it was occupied by Nazi Germany. And all became leading theorists of the Manhattan Project. The other members of the quartet were Edward Teller, Eugene Wigner, and John von Neumann.

From today's perspective, Szilard's intellectual and scientific achievements were the most astonishing of any of the four. As early as 1934, the physicist had filed patents in England about the potential use of neutrons to split the cores of atoms, one of the central ideas of nuclear research. Szilard also possessed remarkable prescience. From the start he recognized the military utility of his ideas and refused to publish them, sharing them instead confidentially with the British admiralty. Even in the early 1930s, he was able to envision the destructive force of atomic fission and knew that such power could never be allowed to fall into Hitler's hands. In 1939, he convinced Albert Einstein to sign a letter to President Roosevelt, urging the United States to develop atomic weapons. This is often described as the moment at which the American nuclear weapons program was born. Ironically, in the course of the Manhattan Project, Szilard changed his mind and criticized the use of such weapons, describing the dropping of atomic bombs on Japan as an act of barbarism.

In the 1950s, this unusually intellectual and politically sensitive man was warning of a new danger: the extermination of the entire human race with cobalt bombs. Szilard was not just a doomsayer. He was also the technical "mastermind" behind this sort of apocalyptic weapon. In a roundtable discussion broadcast on the radio four years before the *Times* published its grim vision of the future, Szilard had spoken about the possibility of constructing a form of military hardware that would contaminate the entire planet. The idea was to surround hydrogen bombs with an element that would suck up the deadly radiation and then transport it around the globe, blown by the four winds. At this point, the hydrogen bomb was only in the developmental stage, and the chances of a manmade apocalypse seemed quite remote. The public paid little interest to Szilard's ideas, which, as was often the case, the brilliant Hungarian meant as a warning. He explained that it would take an enormous number of hydrogen bombs to extinguish all life on earth because radioactivity in a hydrogen bomb gets circulated by carbonates. However, he explained, it would be relatively easy to design a hydrogen bomb that would emit deadly radiation. Most elements that occur naturally turn radioactive when they absorb neutrons. All one needed to do was select an element that absorbed the greatest number of neutrons, and one would be well on one's way toward creating a doomsday weapon. If one chose a radioactive element with a half-life of five years and released it to float around freely in the atmosphere, it would, within that half-decade, be distributed all over the globe and sink to the surface of the earth in the form of "suicide dust." In Szilard's calculations, one would need about fifty tons of neutrons, the amount contained in

500 tons of heavy hydrogen, to kill every man, woman, and child on earth. Szilard went on to speculate about the wider political context:

> What is the practical importance of this? Who would want to kill everybody on earth? ... I think that it has some practical importance, because if either Russia or America prepare H-bombs—and it does not take a very large number to do this and rig it in this manner—you could say that both Russia and America can be invincible. Let us suppose that we have a war and that we are on the verge of winning the war against Russia, after a struggle which perhaps lasts ten years. The Russians and others can say: "You come no farther. You do not invade Europe, and you do not drop ordinary atom bombs on us, or else we will detonate our H-bombs and kill everybody." Faced with such a threat, I do not think that we could go forward. I think that Russia would be invincible. So some practical importance is attached to this fantastic possibility.

Szilard explained how easy it would be to construct this world-suicide weapon, specifically mentioning the element cobalt. Cobalt was both plentiful and inexpensive. All that needed to be done was to build a cobalt case for a conventional, if extremely large, hydrogen bomb. Upon detonation, the element would suck up the deadly radiation like a sponge and transport it in all directions. Moreover, since the purpose of such a weapon was to poison all the earth's atmosphere, it would make no difference where the bomb was detonated. That would obviate the need to consider the size of the apparatus or how it might be transported.

Szilard's warnings fell on deaf ears in 1950, but public interest in the topic was awakened with the detonation of the first hydrogen bomb two years later and then with the *Times* article. Nonetheless, although most of Szilard's contemporaries agreed with him about the feasibility of the cobalt bomb, opinions differed about the Hungarian's strategic thoughts. The historian of science P.D. Smith has collected some of the more interesting reactions in his book *Doomsday Men*. Part of the public reacted hysterically—to some degree with good reason. The Australian government, for instance, had long tried to pressure the British military into forgoing any atomic tests using cobalt bombs, but in 1957 Her Majesty's nuclear physicists had gone ahead and detonated a one-kiloton nuclear device that contained cobalt. The official reason was so the scientists could better track the fallout. There had also been an unsettling incident three years earlier when Italian customs officials seized nine tons of high-grade cobalt in a truck on the Italian-Swiss border. It was unclear where the material was headed, but authorities suspected it was bound for the Warsaw Pact. The Soviet military magazine *Red Star* dismissed the incident, saying that Western worries about the cobalt bomb were exaggerated and encouraged by "imperialists" to keep people in a state of panic. Nonetheless, at roughly the same time, the man who discovered atomic fission, Otto Hahn, issued a warning from West Germany. Human beings, the aging physicist predicted, would soon be capable of destroying the entire world.

In 1957, as the fear of the cobalt bomb was reaching its peak, British-Australian novelist Nevil Shute published his novel *On the Beach*. It centers around the last remaining human survivors of World War III, waiting for the arrival of a deadly radioactive cloud in Southern Australia. The protagonists know they have no way of avoiding their fate and can only watch, helplessly, as death approaches—the captain of a submarine has reported back about a voyage to North America, where he found only scorched earth. In 1959, this apocalyptic saga was made into a film starring Gregory Peck, Ava Gardner, and Anthony Perkins. It was an instant blockbuster. The subject matter fascinated people in Cold War America and was a talking point in the media for weeks.

Only one week before the onset of Cuban Missile Crisis, as humanity came closer than ever to a "hot" nuclear war, *Popular Science* magazine poured more oil on the fire. A September 1962 article with the headline "Man's Last Blast" laid out a series of terrifying end-of-the-world scenarios. In it, author Michael Mann compared the weapons being developed in the laboratories of the East and West to a vial of nitroglycerine held aloft by a robber threatening to blow a bank to smithereens if the money wasn't handed over. All indications were, Mann claimed, that both sides of the Cold War were constructing monstrous doomsday weapons. With a certain amount of unmistakable perverse glee, *Popular Science* examined the technological variations for ending life as we know it. One possibility was to radically alter the earth's climate—although this was deemed a worry for the future. Alternatively, military generals could try to reduce planet's surface to cinders by creating an atomic firestorm—according to Mann, this was a "realistic" option. With only fifty tons of heavy water ("at 30 dollars a pound"), scientists could build a 1,000-megaton bomb with an explosive force of one million tons of TNT. If such a bomb were detonated at an elevation of sixteen kilometers, it would generate such enormous heat that it would act like a miniature sun, instantly destroying an area of 460,000 square kilometers. Bombs that could be detonated in bad weather were especially deadly since cloud cover would reflect the heat earthward. A third possibility that Mann put forward was a cobalt bomb with an additional casing of sodium, an element that likewise absorbs neutrons but that, owing to its shorter half-life, would emit more radiation and thus be a quicker, more "humane" means of exterminating humanity. With its longer-term effects, cobalt would then be used to contaminate those who tried to hide óut in antiatomic bunkers. Submarines could bring gigantic bombs of this type to strategic points in the world's oceans, where the detonations would cause thirty-meter tsunamis and obliterate coastal cities.

Edward Teller was incensed at reports like this that stirred up panic, and he wrote that cobalt bombs did not exist and would be militarily useless since they would not distinguish between friend and foe and would take years to unfold their lethal effects. But reassurance of this sort probably did little to calm frayed nerves, coming as they did from the father of the hydrogen bomb. Even if we can be reasonably sure that the cobalt bomb was never built, horrific fantasies about one fit in well with the generally apocalyptic mood of the period.

Specters of the doomsday weapon disappeared with the end of the Cold War, but they reemerged, unexpectedly and in a new guise, in 1993. Once again, the spark was a major article in the *New York Times*. The object of fear this time around was not a cobalt bomb, but rather a complex Soviet defense system for automatically triggered reprisals in case of an atomic first strike by the United States. The crux here was the mechanistic nature of the system, which was allegedly motivated by fears that Washington could take out the Soviet leadership, so that they could no longer order the deployment of Russian nuclear weapons. "Perimetr," as the system was allegedly code-named, would automatically unleash retaliation if (1) the general staff in the USSR activated it; (2) hidden sensors reported atomic weapons striking Soviet cities and military installations and (3) contact with the Soviet general staff was broken off, in which case it was assumed that the military and political leadership were dead. If all these criteria were met, "Perimetr" would empower an on-duty officer stationed deep within a mountain in the Urals to launch communications rockets. They would then broadcast attack orders to all Soviet submarines, missiles, and bombers. In this way, America and Western Europe could be destroyed by a machine serviced only by a lone human being. Also known as "Dead Hand," "Perimetr" seemed to have been taken directly from the film *Dr. Strangelove, or How I Learned to Stop Worrying and Love the Bomb*. In Stanley Kubrick's 1964 satiric masterpiece, the Soviet Union retaliates for an atomic attack initiated by an insane U.S. Air Force general by launching hundreds of underground nuclear missiles. In the 1980s, reality had ostensibly caught up with fiction.

The man upon whom the "Perimetr" report in the *Times* was based was a former American launch-control officer who had been in charge of dozens of nuclear-armed Minuteman ICBMs. After being discharged from the military, Bruce Blair earned a doctorate and quickly built up a reputation as a theoretician and strategist. In the short phase after the breakdown of the Soviet Union when the interests of the two superpowers converged, Blair served as an advisor for both sides. Having been initiated into "Looking Glass," an airborne defense system with which the U.S. president would have coordinated a response in the event of a Soviet first strike, Blair was already familiar with a military mechanism not unlike "Perimetr." He also had firsthand knowledge of the strengths and weaknesses of America's nuclear arsenal and used the diplomatic thaw to improve security standards in both the East and West. It was in his capacity as advisor to the new Russian state, with all its early turbulence, that he claimed to have been fed information about "Dead Hand." Former members of Soviet missile divisions allegedly told him of how seismic disturbances and radioactivity would activate the system, which would then communicate with hidden sensors using low-frequency signals.

Blair's primary source was Colonel Valeri E. Yarynych, a thirty-year member of the Soviet missile divisions and general staff. Yarynych claimed not just to know about "Dead Hand," but to have helped develop it. The existence of such a system was also confirmed by several former high-ranking Soviet functionaries, including Vitaly Leonidovich Katayev, who worked in the Soviet Ministry of Defense from 1967 to 1985, and Former Deputy Chief of Staff of Strategic Rocket Forces Varfolomei Vladimirovich Korobushin. Respectable sources thus corroborated that "Perimetr" was not just a figment of Blair's imagination. On the other hand, the *New York Times* article also cited several top-level cabinet members from the Reagan and Bush Sr. administrations, including later Defense Secretary Robert Gates, who doubted the existence of "Dead Hand." They all claimed to know nothing of the Soviet doomsday system. So what was the truth?

To answer this question, we first must consider whether something like "Perimetr" would have made any sense. Since the U.S. government apparently knew nothing about it, it seemed to be useless as a deterrent. According to Yarynych, speaking to *Wired* magazine, "Perimetr" was invented as a response to the aggressive stance of the Reagan Administration toward the Soviet Union. The deployment of Pershing II missiles in West Germany in 1983 had reduced the amount of advance warning the Soviets would have had about a nuclear attack to less than a quarter of a hour. In other words, Moscow would have about fifteen minutes to decide whether an incident represented a American first strike or a false alarm. Moreover, the political and military leadership would have had to consult about whether to counterattack and issue corresponding orders to Soviet forces. An all-too-familiar example shows how unrealistic it is to expect leaders to reach sensible decisions under that sort of pressure. On September 11, 2001, when George W. Bush was informed that a second plane had crashed into the World Trade Center, he was reading aloud to schoolchildren from the book *The Pet Goat*. It took him seven minutes just to stand up and leave the classroom so he could form his own views of the situation.

It's hard to blame individuals for being overwhelmed by events of massive historic proportions. But shocked paralysis was not a state anyone could afford during the Cold War, especially not when one's enemies were capable of launching a devastating first strike. "Perimetr" solved this problem, buying the Soviet leadership a bit more time. As soon as radar stations reported an American missile attack, the general staff could activate the system and analyze the situation with cool heads. In case of a false alarm, not that uncommon during the Cold War, leaders could have simply deactivated "Perimetr." However, if the attack turned out to be real, and it destroyed the Kremlin and killed the Soviet leadership, the system would have taken over and ordered nuclear retaliation. Thus, according to Yarynych, "Perimetr" was intended to deter the Soviets themselves from making premature and potentially catastrophic decisions under extreme pressure. There were a number of almost fatal near-misses in 1983. The Soviet leadership initially mistook large-scale NATO maneuvers codenamed "Able Archer" for attack preparations, and a Cosmos spy satellite also once erroneously reported that the USSR was under nuclear attack. A single Soviet officer, Air Force Lieutenant Colonel Stanislav Petrov, duty officer at the command center for the Oko nuclear early-warning system, prevented what could have turned into a nuclear exchange. Therefore, according to the perverse logic of the Cold War, "Perimetr" did indeed make sense.

What does appear suspicious is a contradiction between the statements of those who claimed to have witnessed or knew about "Dead Hand." The former Soviet officers do not agree about when "Perimetr" went operational. Yarynych speaks of 1984, while Katayev talks about the early 1980s, and definitely by 1981. It is also unclear how "Dead Hand" was integrated into the Soviet Union's total nuclear arsenal. In the *Times*, Blair writes that "Perimetr" rockets would broadcast attack orders directly to automatically firing missiles, but a U.S. military observer who was allowed to tour the local nuclear base of Zaratov in the Russian Federation after the breakdown of the Soviet Union tells a different story. As a U.S. military observer who toured the Soviet nuclear facilities contended: "They use a process of three person control. And Gen. Valynkin made it a point that at his national facilities, it's four person control . . . We . . . tend to use technology a heck of a lot more than the Russians do. They're still very manpower intensive, but that's working for them."

It seems, then, that Soviet security architecture was intentionally built around people rather than technology. The military observer's report mentions no indications that the system had been automated, and that would not have fit in the general philosophy of the Soviet military. Soviet nuclear weaponry could only by activated by using several keys distributed to various figures of responsibility representing various organs of state (the military, the KGB, the Communist Party). Mutual surveillance was the rule, both among the bearers of the keys and within individual hierarchies and organizations. The U.S. observer was impressed with the Soviet system and praised it as very secure. At the other end of the scale, by the way, was Great Britain, where—according to a BBC report—nuclear weapons were secured with a single, conventional bicycle lock up until the end of the 1990s.

Nonetheless, in view of the perceived heightened threat from the West in the 1980s, it is hardly impossible that the Soviet Union might have pondered a departure from standard safety procedures. "Perimetr" may well have existed as a concept well along in the developmental stage. But it is unlikely that it was ever constructed—and that it is now slumbering deep inside some mountain in the Urals.

Former Deputy Director of the Central Scientific Research Institute for General Machine Building Viktor Surikov has claimed that he helped design "Perimetr," but that the system was never built. Surikov had thirty years of experience constructing missile, satellite, and communication systems. If anyone would have known specific details about "Dead Hand," Surikov was that man. In 1993, he was interviewed by an American security expert who reported:

> Dr. Surikov responded that he and his subordinates had designed the system to include various sensors— seismic, light, and radiation—to launch the command missiles in the event that the leadership were dead or unable to communicate. He continued that he briefed the concept to his chief, then Institute Director Mozzhorin, and to Baklanov, then the Central Committee Secretary for military industry. Both accepted

and approved the concept. The design finally was rejected by Marshal of the Soviet Union Akhromeev
[evidently when he was Chief of the General Staff, i.e.
after September 6, 1984] on the recommendation of a
trusted advisor and general officer, General-Colonel
Korobushin [the officer who "revealed" the existence
of the system to me months earlier]. As a result of this
rejection, the "Dead Hand" trigger mechanism was
"never realized."

CHAPTER SEVEN
Flying Reactors

There are places where radioactive substances have no business being. One of them is space. That may sound like science fiction, but it is, in fact, everyday reality. Solar-powered sails generate insufficient power to keep many types of satellites and space probes running. The Cassini-Huygens spacecraft, for instance, which arrived at Saturn in 2004, used atomic batteries that generated energy by harnessing the decay of plutonium 238 atoms. The advantages of nuclear batteries for space travel are obvious. They are extremely durable and capable of reliably generating the amount of energy required for long trips through outer space.

By the end of the 1960s, leaders and technicians in the Soviet Union, the first country to develop satellite technology, began to see even the most powerful nuclear batteries as insufficient to their needs and focused on building reactors to power new satellites. Miniature atomic power plants with uranium 235 fuel were used to power RORSAT military surveillance satellites. These mechanical spies watched over enemy naval movements from low-earth orbit. At the end of their lifespan, they were simply blasted into higher orbits. The idea was that their nuclear fuel could continue to decay, for centuries, without harming anyone. In the years before 1998, dozens of RORSATs ended up in what amounted to a space junkyard.

The builders of these satellites didn't want to acknowledge that eventually something would go wrong with the nuclear reactors orbiting the earth. They had equipped the satellites with emergency rockets to offload the reactors if necessary, and they claimed their system was foolproof. But in November 1977, they found out just how real a possibility a catastrophe could be. Cosmos 954, a fourteen-meter-long RORSAT with forty-five kilograms of uranium on board, suddenly began to lose altitude when technicians tried to send it into higher orbit. For reason that remained unknown, the emergency system failed. Attempts to detach the reactor failed. As Cosmos 954 approached closer and closer, there were frantic debates on the surface of the earth. On November 1977, the Soviets informed the U.S. State Department about the situation. A staff of American experts was set up, and they were able to predict the day on which the satellite would re-enter the earth's atmosphere: January 24, 1978. But no one knew where the live nuclear reactor would crash. One of the experts involved described the enterprise as "playing night baseball with the lights out."

Over the course of January, the press got wind of the impending Cosmos catastrophe. Newspapers immediately began speculating about what was in store for humanity should the reactor crash into New York, Paris, or Tokyo. Journalists made the situation sound like a gigantic game of Russian roulette. While the Soviets assured the world that there would be no meltdown, in the middle of the Cold War no one knew whether the Kremlin leadership could be taken at its word.

By the time Cosmos 954 began to plunge toward North America during its 2,089th orbit, people's nerves were completely frayed, but the world got lucky. *Time* magazine calculated that if the satellite had orbited the earth one more time it could have crashed in New York City at rush hour. As it was, Cosmos 954 came to earth in a sparsely populated area of northwestern Canada. It was unclear where the highly radioactive wreckage was, but there were some indications as to its location. Eyewitnesses reported seeing Cosmos 954 in the morning sky above the Northwest Territories. It had crashed to earth like a glowing meteorite, they said. Experts surmised that wreckage was scattered between Great Slave Lake and Baker Lake.

The weeks and months that followed saw one of the most expensive and complex salvage operations in the history of the Space Age. The name of this logistical feat was "Operation Morning Light," and it took place under the most unfortunate circumstances imaginable. The weather was miserable—Cosmos 954 had been inconsiderate enough to crash during the middle of the Arctic winter. Late-January temperatures in northern Canada often dipped well below minus-30 Celsius. Moreover, the area that reconnaissance teams would have to cover in search of wreckage was far from any human settlement. From the onset, it was clear that only the military could carry out a mission of this scope and urgency.

The Royal Canadian Air Force base at Namao, near Edmonton, Alberta, was designated as mission headquarters. No sooner had Cosmos 954 come down than expert Bob Grasty arrived from the Geological Survey of Canada (GSC). The GSC possessed a state-of-the-art spectrometer capable of detecting radiation from Cosmos 954's reactor from the air. But even the experienced Grasty and his colleagues were dumbfounded by what they saw in Namao:

They had arrived complete with all their gear in two C-141 military cargo aircraft, which had brought all the paraphernalia and the one hundred and twenty people that came with it. The C-141 cargo plane was a behemoth about the size of a 747 airliner. The equipment that it disgorged included not only a complete mobile communications centre, with powerful relay transmitters to communicate with their home base, but two helicopters with gamma ray spectrometry systems already installed, as well as several other spectrometry systems ready to be installed in any available aircraft. Also included were two complete data processing units in what came to be known as the "bread vans," because they were the type of small delivery vans typically used by bakeries.

This equipment was the property of a group from the Nuclear Emergency Search Team (NEST) that had been flown in from Las Vegas. Their usual job, among other things, was to fly sorties over areas used for underground nuclear tests and search the ground for radioactive leaks.

Grasty and his colleague Quentin Bristow brought their own high-tech gear and joined the gigantic search operation as its only team of Canadian geologists. In the beginning, they had an advantage over the other searchers—the only winterproof spectrometer. The Americans' measuring devices were designed for use in the Nevada deserts and were no match for the minus-40-degree chill of Great Slave Lake. The NEST teams solved this problem by loading their spectrometers into fully heated Hercules military transport aircraft, which then flew zigzags across the entire search area. Days went by, but searchers were unable to find even the slightest indication of abnormal radiation. With the Canadian press and political establishment breathing down their necks, mission command was getting visibly nervous.

While the Hercules aircraft were making their rounds in the Great White North, six outdoorsmen were dogsledding some 640 kilometers east of Great Slave Lake. The men were scientists who were retracing the journeys of legendary nature researcher John Hornby through the wilderness. The routes they took were quite challenging even for a team equipped with all the latest technology. In 1926, Hornby and two fellow scientists had made camp in this remote region. In an age in which hardly anyone was interested in the Canadian wilderness, the trio of researchers intended to observe nature there for an entire year. Hornby had a lot of experience with Polar regions, but the expedition to Theron River ended in tragedy. The men failed to run into the herds of caribou that migrate south every fall. Hornby had planned to shoot a number of caribou and dry the meat for the winter. There was no time to return to civilization before the snows came. At some point in early 1927, Hornby and his colleagues' supplies were exhausted, and all three starved to death.

The group led by John Mordhorst and Mike Mobley who followed in Hornby's footsteps a half-century later would have seen seen the three researchers' graves, simple wooden crosses in a clearing in a pine forest. The only information they had about the search for the Cosmos 954 would have been what they heard on short-wave radio. Like Hornby, they had been caught off-guard by an early winter snow. Mordhorst, Mobley, and rest, however, had been able to take shelter in a well-supplied log cabin. Using that as a base, they were able to make short exploratory trips throughout the winter. On one of their trips with dogsleds, Mobley and Mordhorst discovered a crater in the snow with a strange bit of metal sticking out of it, almost like antlers. The two immediately realized that this could only be a part of the crashed nuclear satellite. When they returned to their cabin, they contacted the Canadian authorities and told them what they had found. Before long, a Chinook transport helicopter was hovering overhead. The entire expedition team, including the sled dogs, was evacuated. Mobley and Mordhorst had to submit to a thorough medical examination. There were fears that they could have radiation poisoning. In fact, they had only been exposed to a small amount of radiation, roughly the equivalent of two chest X-rays.

When the press got wind of the satellite wreckage, reporters from the town of Yellowknife tried to make their way to Theron River, defying both the icy temperatures and the danger of potentially lethal radiation. The Canadian military sent in paratroopers to cordon off the area. But it soon emerged that the piece of the satellite discovered by Mordhorst and Mobley was not connected, under the snow, with the reactor. There was still no trace of it. This wasn't the first disappointment that the authorities had suffered. A short time earlier, there had been another false alarm after a surveillance flight turned up a radioactive object. The Canadian defense minister was convinced that this had to be either a piece of the Cosmos wreckage or "the world's largest uranium mine." It turned out, though, that one of the sensors on the plane was picking up the onboard electronics. This embarrassing mistake only increased the pressure on all concerned to produce some results.

It took a lot of expeditions before the searchers achieved a breakthrough. Ironically, it was the small team led by Grasty and Bristow who came back with suspicious readings. Their sensors had registered some radiation near a tiny village called

Snowdrift on an island at the east end of Great Slave Lake. Ice there was contaminated with lanthium 140, a product of atomic decay that could only have come from the reactor. Helicopters immediately swarmed the area, finding a number of parts of Cosmos 954. They marked the site with red dye, and a ground team in protective clothing was sent in to collect the wreckage. Many parts of the reactor were so radioactive that experts could only pick them up with long tongs from behind lead shields. High winds also hampered cleanup operations. Nonetheless, despite all the obstacles, the search was declared officially over on February 4, 1978. Five parts of the satellite had been discovered in the icy desolation. All of them were highly radioactive.

The costs of the mission were horrendous, although Canada only demanded the modest sum of six million Canadian dollars as compensation from the Soviet Union. After a protracted series of negotiations, the Kremlin reluctantly agreed to pay half of that sum.

This was by no means the end of the drama surrounding nuclear-powered satellites. On February 7, 1983, another atomic-driven Soviet spy satellite, Cosmos 1402, failed to achieve its proper orbit. The satellite plunged into the Indian Ocean while its radioactive core came to earth in the South Atlantic. Five years later, the Soviets lost control of Cosmos 1900. The official word was that the satellite had been able to send itself into a higher orbit. In 1996, a Russian space probe intended for Mars made headlines when it burned up over the Pacific, with plutonium on board.

What's going on above our heads continues to be a cause of concern. There are still dozens of RORSATs from the Cold War circling in what are known as "disposal orbits," and space around planet Earth is getting tight. In February 2009, an iridium communications satellite collided with a decommissioned Russian military satellite. Both were destroyed on impact. And engineers continue to use radioactive material as a power source—for instance, in the space vehicle for the Mars Science Laboratory, which successfully landed the rover Curiosity on the Red Planet.

Fortunately not all of the ideas imaginative engineers came up were put into practice. Thankfully, atomic-bomb-propelled Orion research rockets never made it off the drawing board, and neither did the streamlined Ford Nucleon with a nuclear reactor in its trunk. Another risky idea, however, did get a bit closer to becoming reality: airplanes powered by nuclear-driven jet engines. Both the United States and the Soviet Union worked on this kind of aircraft. In August 1954, the German news weekly *Der Spiegel* described the less-than-trustworthy-sounding propulsion system as "less a motor than a kind of atomic oven that generates heat by smashing atomic nuclei." The atomic oven was basically "nothing other than a controlled atomic bomb whose explosion takes place unimaginably slowly." The gizmo whose heat would be used to compress air for the jets was a "fast" mini-reactor. It had no chemical moderator like graphite that would put the brakes on the chain reaction as in a nuclear power plant. In 1948, having seen the very first designs for such an aircraft, *Time* magazine enthused: "A 'nuclear-powered' plane could fly on & on, round & round the earth. It could fly at its top speed all the time, and land with the same weight (and about the same amount of reserve fuel) that it took off with."

In the imagination of engineers-cum-Cold-Warriors, the atomic plane had to be supersonic, and it was conceived right from the start as a military project. That was why its development got a lot further than, for instance, the purely civilian atomic locomotive. In the test of strength between the super-powers, neither side wanted to shy away—least of all when it came to the bombers that would circle the globe like mad insects, ready at any time to deliver the fatal sting.

The background to the lunatic idea of nuclear airplanes was complex and political. The Americans believed that they had fallen behind the Soviets in the development of long-range aircraft. At their 1954 May Day parade, the Soviet Union had unveiled a brand-new, conventionally driven stratospheric bomber called the Bison. The imposing aircraft was allegedly capable of reaching every city in the United States and dropping atomic bombs. (It later emerged that the Bison did not have the range to make a bombing raid in the United States and successfully return to the USSR.) While the U.S. Air Force did possess comparable jets, developers were still struggling to work out a whole series of bugs. As a result, worries arose about a "bomber gap" that might give the USSR a decisive aerial advantage over the United States. The Air Force frantically poured its energy into long-range bombers. Thousands were built, and the costs, converted into today's dollars, sky-rocketed into the billions. At the same time, Air Force strategists fretted that they were missing the next big innovative leap—the atomic jet. Corresponding plans had been slumbering in desk drawers since the 1940s. They had been drawn up by Andrew Kalitinsky, an atomic engineer who worked for the Pentagon and the Fairchild aviation corporation. Various airplane manufacturers had submitted blueprints with variations on the idea. Lockheed suggested building a bomber that resembled an oversized business jet. De Havilland came up with rocketlike machines that would streak around the earth at twice the speed of sound and at an altitude of 18,000 meters, stuffed to capacity with hydrogen bombs.

In the end, the masterminds of the atomic jet had to admit that the laws of nature put limits on madness. Kalitinsky suspected that reactor-driven motors would end up killing the plane crews. "The radiation intensities encountered in nuclear reactors must be reduced by factors of many billion before they are safe for the human organism," Kalitinsky told *Time* in 1948.

But in the mid-1950s, U.S. military leaders weren't necessarily put off by such practical considerations. Two years after the shock of the "bomber gap," a Convair B36 "Peacemaker" took off with a nuclear reactor on board. The plane itself, however, was conventionally propelled—the flight was only aimed at testing how the reactor would perform while airborne. The plane's crew was protected by a twelve-ton shield of lead and large, radiation-absorbing tanks of water. The cockpit window was made of thirty-centimeter-thick leaded glass. All told, this specially equipped plane made forty-seven flights.

Kalitinsky had optimistically assumed that within a few years, lead could be replaced by ultra-light materials equally suitable for acting as a shield against radiation. But his hopes were dashed, and with them the dream of a long-range nuclear airplane. The Soviets, who were experimenting with a similar propulsion system using a modified Tupelev, came to likewise sobering conclusions about the feasibility of such an aircraft. Militarily, the advent of ICBMs made the idea of a nuclear-powered bomber plane obsolete anyway. This was fortunate, since the plane's developers had no answer to the question of what would happen should one of the flying reactors ever crash. The most sensible concept, if the word "sensible" can even be used in this context, was an idea of the De Havilland Aircraft Company. The company's design foresaw a gigantic atomic amphibious plane that could take off and land on water. In the worst-case scenario, a defective nuclear plane would crash into a body of water and sink, together with its radioactive propulsion system.

Such Jules Verne–esque visions never proceeded beyond the early conceptual phase. The situation was different with "Pluto," an ultra-fast cruise missile that could circle the globe several times, propelled by a high-performance reactor. The U.S. military built a functioning prototype of the engine, but the missile system, which would have been able to deliver atomic warheads to several targets, never went into serial production. Washington didn't want to provoke the Soviets by building such a hellish weapon. On this occasion at least, the military leadership refused to add further horrors to the dystopia they had already created.

Edward Teller, the "father of the hydrogen bomb," on a PR tour
through Alaska. Teller wanted to use five nuclear bombs to build
a new port in the far north.

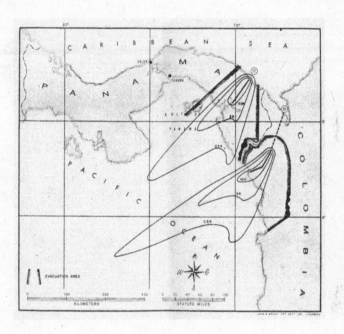

Plans were drawn up to use "atomic earthmoving" to create
a seventy-three-kilometer canal in southern Panama. Alternate
routes were also planned. This illustration shows the probable
drift of fallout in two variations.

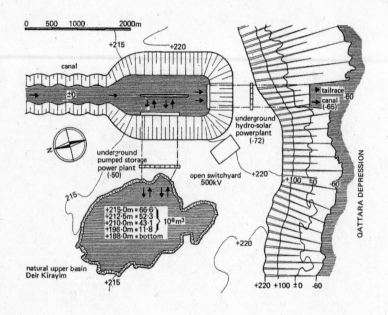

German engineer and professor Friedrich Bassler planned to excavate a canal in Egypt that would connect the Qattara Depression with the Mediterranean Sea. Two hundred hydrogen bombs would have done the excavating.

Three ways to end the world: change the climate drastically,

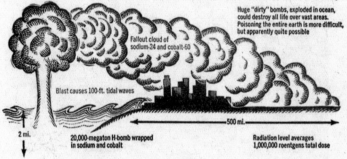

Fallout cloud of sodium-24 and cobalt-60

Huge "dirty" bombs, exploded in ocean, could destroy all life over vast areas. Poisoning the entire earth is more difficult, but apparently quite possible

Blast causes 100-ft. tidal waves

500 mi.

2 mi.

20,000-megaton H-bomb wrapped in sodium and cobalt

Radiation level averages 1,000,000 roentgens total dose

In 1962, the magazine *Popular Science* discussed possible effects of a cobalt bomb. If one was detonated in the ocean, it would cause not only a deadly radioactive cloud, but a tsunami.

The logo of "Operation Morning Light," the mission to recover the remains of the atomic satellite Cosmos 954 in the icy north of Canada.

A model of the Ford Nucleon—the car with a reactor in its trunk.

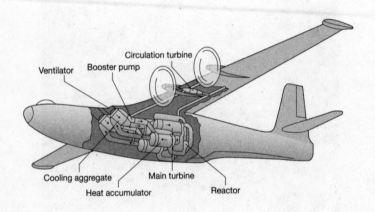

A sketch of the De Havilland amphibious nuclear plane.

Exploratory troops, wearing gas masks, during the British nuclear test
Operation Hurricane. Whether all the soldiers involved were given
adequate protection remains disputed.

Dummies were used in Maralinga, Australia, to simulate the
effects of an atomic explosion on human beings.

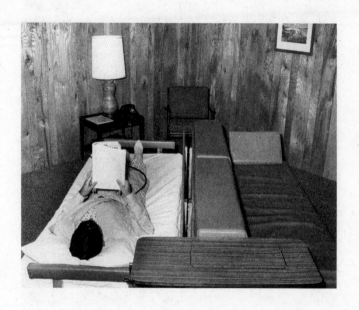

Experiments were carried out in the United States during
the 1940s and '50s to determine the effects of radiation on the
human organism. The machine to bombard the subject's body
looked a lot like a room at a Holiday Inn.

An exhausted Jens Zinglersen takes a break during rescue-and-recovery operations after an American B-52 carrying four hydrogen bombs crashed in Greenland.

The claw of the STAR III submarine searches for bomb
parts in the water off Saunders Island.

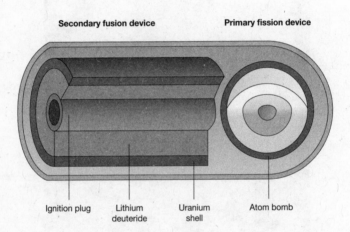

Secondary fusion device **Primary fission device**

Ignition plug Lithium deuteride Uranium shell Atom bomb

Cross-section of a hydrogen bomb.

Admiral William H.P. Blandy and his wife cut a cake to
celebrate Operation Crossroads, while Admiral Frank J. Lawry
looks on. Because of these nuclear tests, the Bikini Atoll remains
unfit for human habitation to this day.

CHAPTER EIGHT
How Safe Is Safe?

Monsignore Luc Gillon was a jolly friar, a jovial and highly intelligent chain-smoker with rosy cheeks and a thinning crown of hair. Gillon also possessed a coolly analytical mind, graduating at the top of his class from Princeton with a degree in nuclear physics. In the 1950s he went to Congo, which was then still a Belgian colony, as a missionary and was soon able to realize two of his lifelong dreams. In 1954, he presided over the birth of the Lovanium, the first university in central Africa. Today, Gillon is considered the spiritual father to a whole generation of Congolese intellectuals who graduated from the institution. Gillon also tirelessly lobbied for another major project: the construction of a nuclear reactor for research in the heart of Léopoldville, today's Kinshasa. History may ultimately remember Gillon as the creator of the world's most insecure nuclear reactor.

In the early 1950s, Gillon started his petitions to the Belgian government, and he also advised top officials as to how to proceed with the project. The government, he suggested, should demand American nuclear technology as "free-issue equipment." His argument was that since the uranium ore used in the atomic bomb that the United States dropped on Hiroshima came from Belgian Congo, the Americans owed the colony a favor. Astonishingly, this reasoning was accepted, and in 1957, as part of the Atoms for Peace program, Washington had the components for a TRIGA Mark 1 reactor delivered to Congo. This compact facility was made by a company called General Atomic and was installed by on-site Belgian technicians. It went operational in 1959. One year later, Gillon posed for a *Life* magazine photo in front of the reactor's heavy-water pool. The West hailed the reactor as a symbol of progress in Africa. But Washington's gift was hardly as selfless as it seemed. The United States hoped that the TRIGA Mark 1 would curry favor among the elites of the colony, which possessed vast natural resources. Indeed, the reactor was a point of pride for Congo, which achieved independence from Belgium in 1960. But instead of becoming a beacon of peace and freedom, the young nation quickly descended into civil war and chaos.

A decade later, the charismatic and ruthless Mobutu Sese Seko emerged victorious from the fighting and appointed himself president. The research reactor in the middle of the capital was a useful vehicle for the dictator to use to boost the prestige of his corrupt regime, and in 1972 the aging facility was replaced with the much more powerful TRIGA Mark 2. Scientists didn't do anything particularly spectacular with this prestigious reactor, using it to produce radioactive isotopes for medical equipment and to irradiate seeds. The institute where the reactor was kept was also anything but state of the art. The reactor building had neither telephones nor reliable electricity. Calculations were made on blackboards. Then at some point during the 1970s, Mobutu lost interest in the reactor, and no one got especially upset when two seventy-centimeter-long backup fuel rods disappeared from a storage pool. The later director for the department of nuclear physics at the university,

Professor Félix Malo Wa Kalenga, surmised that his predecessor had simply lent a visitor a ring of keys without noticing that one of them was to the reactor. When journalist Michela Wrong interviewed Kalenga about lax security standards at the facility, he pretended to be deaf and unable to understand her questions.

Twenty years later, one of the fuel rods suddenly turned up in Italy. Smugglers offered the uranium, which was at least 20-percent enriched, to what they thought were buyers from the Middle East—in reality agents of the Italian police's Central Investigating Service, SCICO. As a result of "Operation Gamma," authorities apprehended twelve people, all of them members of the mafia, but none provided any information concerning the whereabouts of the other fuel rod. An IAEA expert has said that it's probably somewhere in the Congolese forests. More pessimistic observers think that it has likely already fallen into the wrong hands. Reassuringly, the uranium contained by the rod is nowhere near sufficient enough to generate enough fissible material for a bomb, which would require far greater amounts of enriched material.

The National Nuclear Security Agency, which is part of the Department of Energy, says that by 2013 all of the "most sensitive" nuclear material around the world will have been secured. At the top of the list is the highly enriched uranium used in civilian research reactors. In contrast to military installations, such facilities represent soft targets. And while it's difficult to convert the sort of material used in research reactors into fuel for bombs, it's not impossible.

Those interested in what happened to the Congolese research reactor after Mobutu's downfall and death in 1997 can see for themselves by visiting the decrepit building complex in the Southern Kinshasa neighborhood of Lema. Security is relaxed. A few padlocks are the only things restricting access to the buildings. There's no need to find a hole in the fence to crawl through. Usually there's no guard in the gatekeeper's house.

Since 1997, more fuel rods have apparently gone missing. In March 2007, one of the nuclear center's directors was arrested in Kinshasa. He was accused of selling an "important quantity" of uranium material over the years on the black market. But it's nearly impossible to determine what is true and what is malicious rumor.

But it's time, and not thieves, that is gnawing away most at the TRIGA Mark 2. In 2001, a reporter dared to make his way into the interior of the reactor. Neon tubes crackled above the sunken pool, the walls of the decrepit building were covered in mold, and garbage was floating atop the brackish water. The visibly nervous technical director of the facility, armed with a Geiger counter, hastened to tell the reporter that they probably shouldn't spend too much time there.

If not all of the radioactive material in the Congo reactor has been stolen, then nine million residents of Kinshasa are sitting on a time bomb. The building housing the reactor was built on a hill with an acute potential for landslides, and, while administrators claim the facility is safe, an entire wall suddenly collapsed in 2000. What's more, the complex was once hit by a small explosive, probably a wayward rocket-launched grenade. If radioactive emissions got into the city's water supply, the resulting catastrophe would be beyond repair. After the 2011 Fukushima disaster, the commissioner of the Kinshasa Regional Center for Nuclear Studies, Vincent Lukanda Muamba, assured the world that the reactor was "idle but safe," adding that operations had been shut down for seven years due to a lack of spare parts. Unlike Fukushima, he told reporters at a press conference on March 17, 2011, Kinshasa was located neither in an earthquake region nor near the sea.

Such blasé statements are not merely a problem. The West, too, has realized atomic dreams that are every bit as misguided as Luc Gillon's nuclear research facility. Westerners have taken nuclear reactors and batteries with them wherever they went, be it to the bottom of the sea or to the peaks of the highest mountains. The United States built a nuclear power plant on Antarctica to supply the McMurdo Research Station with electricity and schlepped a plutonium-based battery up a mountain in the Himalayas to power a weather station. (If we believe an article written by *Rolling Stone* journalist Howard Kohn, the true purpose of the installation was to spy on the Chinese nuclear program on the other side of the mountain.) The ball-shaped battery and its half-kilogram of plutonium were buried in an avalanche. Today, the radioactive battery is probably somewhere on the southern slope of Nanda Devi near the source of the River Ganges. Indian officials deny the story, calling it a cleverly placed media ruse to discredit the United States. In the Cold War, it was seldom possible to separate truth from fiction.

What is beyond doubt is that there is complacency and irresponsibility everywhere, including in Europe, Japan, and the United States. Indeed, in the west, the scale of the nuclear facilities is so much greater that the possiblity of catastrophes is also hugely increased.

The irony is that the early atomic age seemed so full of promise. The new technology was viewed as a potential savior that would solve a host of the world's problems. Initially, nuclear technology seemed to offer a way to defeat Hitler's Germany. Then it became an inexhaustible energy source that would guarantee the prosperity of coming generations. When the first nuclear reactor became operational, humanity took a step into the unknown. The inventors were well attuned to the solemnity of the occasion and the dangers the new age entailed. The first reactor had been conceived within the framework of the Manhattan Project and built according to a design by the brilliant physicist Enrico Fermi. It was kept in a gymnasium at the University of Chicago, a somewhat banal location for a piece of technology that would define a historical period. The reactor's nickname was likewise unpretentious: the Chicago Pile, or CP1 for short. It represented a massive achievement for its constructors. America's entry into the Atomic Age was truly revolutionary. In terms of physics, this was completely uncharted territory. Never before had human beings been able to create tremendous amounts of energy from what seemed like thin air. But the atomic pioneers created a host of ethical problems whose scope no one could have anticipated in 1942. It was a hopeful if uncanny start, and America's nuclear scientists were commensurately nervous at 3:00 p.m. on December 2, 1942, as they commenced the greatest experiment in human history. Fermi had prepared every minute detail before the big day. He and his assistants had pondered security measures for months. This was not only the birth of nuclear power, but nuclear safety. Mechanical systems were in place to automatically shut down the reactor if it started overheating. In addition, an assistant with an axe was stationed next to the pile. In an emergency, he would chop through a rope to lower a master rod into the core of the reactor. As soon as it made contact, nuclear fission would cease. The assistant was nicknamed the

"Safety Cut Rope Axe Man," or SCRAM. Today, the word is still used to describe a rapid emergency shutdown of a nuclear reactor.

In the eventuality that these safety precautions failed, Fermi had posted three further assistants around the pile who were to flood it with a cadmium-saline solution if given the signal. Thankfully, none of the measures was needed. The reactor was booted up at 3:20 p.m.; after about half an hour, Fermi interrupted the chain reaction in controlled fashion. The experiment was a success.

The test in Chicago was very much the child of World War II and was aimed at producing plutonium for a bomb. After Germany and Japan had been defeated, researchers pressed forward with the construction of civilian nuclear power plants in the United States. Nonetheless, military utility always played an important if covert role in the planning of power plants. Nuclear plants inevitably produce plutonium uranium than can be used to make bombs. That is why national programs like the one recently announced by Saudi Arabia to build reactors should always be greeted with skepticism. Civilian and military nuclear programs are Siamese twins.

Historically, the United States was quickest to turn theory into reality. By December 20, 1951, a research reactor in Idaho was already generating electricity. Five years later, the Pentagon unveiled the first atomic submarine. In 1954, the first nuclear power plant went online in the Soviet Union, although it actually consumed more energy than it produced. In 1956, the first truly efficient plant, Calder Hall, began operations in Britain. The Golden Age of nuclear energy appeared to be at hand. An effusive President Eisenhower declared that humanity was standing on the threshold of creating a "new and better earth," and the recently crowned Queen Elizabeth II also spoke of a new world that had opened up thanks to nuclear power.

For a brief moment in history, people truly believed that the planet's energy problems had been solved once and for all. The head of the Atomic Energy Commission boasted that electricity would become so cheap that there would be no need for meters any more. But none of these high hopes was realized. Scientists never found a way to generate electricity using fusion, the process that takes place within the sun, nor did they succeed in developing economical nuclear-fission power plants. Construction costs for such facilities proved disproportionately high compared with other systems for generating power. But there was no going back. The world had invested far too much, in both a monetary and an emotional sense, in the new technology. Politicians felt bound to keep the promises that had raised the public's expectations, and in the background, the military was eager to get its hands on plutonium. Energy companies were forced to swallow horrendous developmental costs.

Firms like General Electric and Westinghouse advised that nuclear plants would have to be built in gigantic formats if nuclear power was to make economic sense. This view was based on drawing analogies between the pressure-water principle of small reactors on submarines and energy factories capable of producing 600, 800, or even 1,000 megawatts. But the concept of lowering costs with greater volume was a dead end. Reactors only got more complex, not more economical. While the reactors used to propel submarines and icebreakers were relatively compact and easy to control, technicians in big power plants had to deal with highly complicated, kilometer-long cooling systems. In a study of the situation nuclear engineers face in the control room, author Stephanie Cooke contrasts them with plumbers. Whereas plumbers can easily check the flame of a gas boiler, she writes, nuclear technicians can't physically get to the source of potential danger, the reactor's core: that means that nuclear technicians, as well trained or experienced as they may be, often feel as though they're being forced to read tea leaves, instead of being the masters of the latest technology.

By the mid-1960s, as construction on a number of colossal plants was well underway, experts began to suspect that a core meltdown in such gigantic facilities would be absolutely catastrophic. Scientists' minds were haunted by the specter of the China Syndrome—the idea that an ultra-hot reactor core could melt the ground on which it stood all the way down to the center of the planet. In 1992, BBC journalist Adam Curtis made an investigative documentary film about the reactions of industrial and political leaders to this nightmare scenario. A safety committee within the AEC recommended modifying reactors to isolate them more effectively from their environs. The industry agreed to some minor concessions but refused for cost reasons to significantly reinforce the containment shells surrounding reactors' cores. Instead, engineers concentrated on modifying cooling systems that were supposed to prevent nuclear meltdown. David Okrent, a leading AEC official who was responsible for licensing the plants, later admitted that his agency had been forced to compromise by General Electric and Westinghouse. According to him, when asked about the safety features of their design, General Electric representatives indicated that "they didn't want to continue selling nuclear reactors if they were going to have to deal with the core-melt problem." Westinghouse put forward something called a "core catcher," but it never provided any evidence that it worked. Okrent recalled: "Neither company was anxious to deal with the problem, obviously ... It was a kind of threat, I think."

Thus, significant sums were spent to build untested systems that were supposed to prevent reactors from overheating. No further measures were instituted for the eventuality that a reactor, despite all the safety precautions, melted down. That avoided costly investments, but the strategy backfired when large-scale nuclear power plants began operating in the 1970s. On March 28, 1979, there was a major reactor accident at the Three Mile Island plant near Harrisburg, Pennsylvania. A defective valve accidentally shut down the plant's cooling system. The engineers responsible for the reactor were caught completely off-guard. No one had any idea what to do in this situation. Neither the makers of the plant nor the company that ran it had devoted sufficient attention to the meltdown scenario. Technicians could only look on helplessly as a giant hydrogen bubble formed within the plant's interior. A gas explosion almost ripped through the plant's containment shell, but the plant workers averted the worst by releasing some of the hydrogen. The shell remained intact. Improvisation and luck prevented large amounts of radioactivity from escaping into the surrounding area.

But soon there was another, likewise unpredictable chain of events, after which it was no longer possible to avert a catastrophe. The name Chernobyl has become a synonym for the dark side of atomic energy. The constellation that led to disaster there bore some similarities to Three Mile Island. Energy policy in the Brezhnev-era Soviet Union focused on building gigantic nuclear power plants, and to keep the exorbitant costs somewhat under control, corners were cut in safety. Just like their colleagues in the West, Soviet nuclear scientists believed that they had a handle on their technology and that a core meltdown was impossible. This proved to be a fatal mistake. One conservative estimate puts the number of deaths caused by the Chernobyl disaster at 4,000. Others sources say as many 30,000 people died.

The problem, both in the West and behind the Iron Curtain, was a lack of imagination. No one was able to picture the worst-case scenario. Indeed, no one wanted to—a pattern that has repeated itself over the years. The proprietors and constructors of the Fukushima nuclear power plant, as well as many Japanese politicians, also had blinders on. The reactors at Fukushima were built according to plans made by General Electric in the 1960s. Safety specifications had taken account of possible earthquakes and tsunamis, but not the fact that severe earthquakes and tsunamis could occur together. The economic viability of the facility took precedence. Thus everyone was caught off-guard. A natural disaster of the sort that was visited on coastal Japan in 2011 may only happen once in a century, but in a region so seismically active, people should have been prepared for anything. Far more minor phenomena than a giant tsunami have been known to cause accidents in nuclear power plants. In 1975, for example, there was a disruption in an American reactor after a technician accidentally set a piece of foam on fire with a candle.

In the wake of Fukushima, people once again began asking: How safe is safe? The latest generation of nuclear power plants is supposedly more reliable than gigantic electricity factories of the 1960s and '70s, but while progress may have been made in averting accidents caused by mechanical defects and human error, serious, new, and hard-to-anticipate risks have arisen in the meantime. In 2010, the Iranian nuclear reactor in Busher was reportedly infected with the Stuxnet computer virus. The Iranian interior minister denied that this had taken place, but Russian scientists warned that viruses could cause a Chernobyl-like disaster there. If the reports are true, nuclear power has become the focus of cyber-warfare—with unknown consequences.

Germany, a country heavily reliant on nuclear power, could become an interesting case study for nations seeking an exit strategy. After decades of political fights, anti-nuclear demonstrations, and squabbles, the country has decided to wind down its production of nuclear power. The last reactor will be taken off the grid in 2022. Today, more than 20 percent of the energy generated by Germany is produced by renewable sources, and this figure is expected to double within the next eight years. While this sends a strong message to other industrialized nations relying on nuclear power and fossil fuels, it doesn't mean that the problems will just disappear. In an age of free international trade in power, Germans of course still call upon nuclear-produced energy from abroad. Many neighboring countries still depend on outmoded nuclear facilities constructed in the 1960s and '70s. And yet even if every country in the world were to shut down its plants, this would leave the problem of how to deal with tons and tons of nuclear waste heaped up in the past. Unfortunately, in our energy-hungry world, easy answers are hard to come by.

CHAPTER NINE

Atomic Australia

On the first day of creation, the ground split and the ancestors climbed from the cracks. Some took the form of men, others the form of animals. They wandered off in all directions across the globe, singing, playing, hunting, and creating all earthly things. When their work was done, the era of Alcheringa—Dreamtime—was done. The ancestors froze in their tracks, as lizards became mountains, ants became cliffs, and bats became dark caves. The places where these formations can still be seen are sacred to Australia's Aboriginal peoples. For 40,000 years in Australia, nomadic tribes have been following in their ancestors' footsteps, which crisscross the red continent like a fine mesh. In seasonal cycles, Aborigines set off on foot, sometimes traversing thousands of kilometers. They pause at the holy sites along the way during "walkabouts" and gather for ceremonies to summon up the power of souls alive in such places.

In 1957, a young woman named Eddie Milpuddie was on such a walkabout with her family. Her journey had commenced two years previously in the Everard Ranges in Central Australia, and the route to her final destination, the Ooldea settlement in the south, took her straight through the outback. It's hard to imagine a more hostile environment. Temperatures in summer exceed fifty degrees Celsius, while in winter they drop below freezing. There's hardly any vegetation. Only a few trees provide shade, and rivers are nonexistent.

Eddie and her husband and children lived on rabbits and lizards that they caught in traps. They drank from hidden water holes they found using songlines, the lyrics of which list landmarks that the Aborigines use to navigate along the invisible paths of their ancestors. At some point, Eddie and her family entered the land of the Tjarutja. There they found a number of strange signs covered in characters that neither Eddie nor her husband, neither of whom spoke English, could decipher. The signs read: **WARNING**. YOU ARE ENTERING A **RADIO-ACTIVE AREA**.

Eddie and her family pressed on, and when night began to fall, they looked around for a good spot to set up camp, eventually finding a large, circular hollow that offered shelter against the wind. But the spot was not as peaceful as it seemed. When the sun had set completely and the Milpuddies had lain down to sleep, they were startled awake by a sudden din of motors. Seconds later, all-terrain vehicles appeared at the edge of the craters, and soldiers in overalls jumped out. White men yelling in a language Eddie didn't understand forced her and her family to get into one of the vehicles.

After a while, they reached a symmetrically-laid-out settlement of barracks. The "whitefellas" ordered Eddie and her family to stand under a stream of water—it was the first shower of their lives. Once they had dried themselves off, the Milpuddies were examined with a Geiger counter and then sent back under the water. This frightening procedure was repeated four or five times. In the end, the soldiers gave the Milpuddies some clothing and left them in peace. Outside, however, the family's four valuable hunting dogs were shot.

Eddie Milpuddie was pregnant at the time of this incident,

but she lost the baby. Her second child died of a brain tumor at the age of two, and her third was born prematurely and only barely survived. She blamed radioactive contamination in the soil. The crater in which the family had camped out had been created a few months previously by a 1.5-kiloton atomic bomb that had been detonated in the desert. "Marcoo 1" was part of the British nuclear program, which in the 1950s moved from the developmental to the test stage. The Clement Attlee government in London had selected Australia as its test site because it considered it "uninhabited." But no one had considered the fact that an Aboriginal path might lead directly through the test zone.

A few years ago, Lorna Arnold, an official historian of the British Ministry of Defense, justified her government's treatment of Aborigines by arguing that, at the time, the latter had no legal rights—a problem general to 1950s Australian and British society and thus not specific to the atomic tests. Arnold asserted that the greatest damage suffered by the Aborigines from the tests was to their "way of life rather than directly to their health." The fact that the Aborigines' interests in the land were neither registered nor respected, Arnold wrote, was because of "their general situation and was neither new nor peculiar to the weapons trials."

The British military carried out nuclear tests in Australia for almost a decade, with the vast majority of the experiments taking place on Aboriginal lands. The reasons for this were twofold. For strategic reasons, Britain wanted to join the nuclear club whatever the cost. Like many other nations in Western Europe during the Cold War, the United Kingdom felt threatened by the Soviet Union. Her Majesty's military leaders reckoned that about a dozen nuclear weapons would suffice to wipe out Britain's most important cities and industrial sites, and London was not comfortable relying on American nuclear deterrence alone. Yet this logic might have applied elsewhere in Europe. The Netherlands, Spain, Italy, and any other European country could also have used it to justify acquiring nuclear arms. In fact, only France—another former colonial power whose influence was waning—went down the same path as Britain. Imperial pride was thus a major factor in London's decision-making. The nation whose empire had extended across the globe at the start of the twentieth century now feared that it would pale into insignificance, overshadowed by the world's two new superpowers. Neither Great Britain nor *la grande nation* wanted to take a seat at the side table. Nuclear bombs were the key to being taken seriously, to having a say in world affairs. Margaret Gowing, another official British government historian, writes:

> The British decision to make an atomic bomb had "emerged" from a body of general assumptions. It had not been a response to an immediate military threat but rather something fundamentalist and almost instinctive—a feeling that Britain must possess so climacteric a weapon in order to deter an atomically armed enemy, a feeling that Britain as a great power must acquire all major new weapons, a feeling that atomic weapons were a manifestation of the scientific and technological superiority on which Britain's strength, so deficient if measured in sheer numbers of men, must depend.

Britain's nuclear ambitions were nothing new. As early as the 1940s, almost two years before the start of the Manhattan Project, scientists Rudolf Peierls and Otto Frisch had reported to the government in Westminster about the possibility of constructing an atomic bomb. The two men, refugees from Germany and Austria, respectively, calculated that a critical mass of just under a half-kilogram of uranium 235 was needed to unleash a spontaneous nuclear chain reaction. Their estimate was wrong and was later revised upward, but it helped convince the British political leadership of the feasibility of the bomb. Churchill reacted immediately. More quickly than any other contemporary political leader, the prime minister recognized the potential importance of nuclear technology as a military weapon. Only a few months later, a secret organization code-named "Tube Alloys" was feverishly working on a British atomic bomb. Ultimately, though, Britain was too small and too preoccupied with the task of defending itself against German bombardment to be equal to the huge industrial effort and output required to produce weapons-grade uranium. By 1942, the Americans had left their British allies in their wake. One year later, Churchill succeeded in persuading Washington to admit the UK as a junior partner in the American nuclear program, but although Britain sent its best nuclear physicists to Los Alamos, there was no reciprocal transfer of information in the other direction. On the contrary, one year after the end of World War II, the United States Congress passed a law imposing draconian punishments, including the death penalty, upon anyone found guilty of betraying American nuclear secrets to other countries.

Britain's pride had been wounded, so it's hardly surprising that London decided to pursue a nuclear program of its own. In 1950, the United States refused a request by the Attlee government to carry out the first British nuclear test in Nevada. In the search for alternatives, the focus turned to Australia. The continent was only sparsely settled, and the Australian government had extremely cordial relations with Britain. In 1952, Australian Prime Minister Howard Menzies immediately gave a green light to "Hurricane," a test that would see a plutonium bomb detonated in the Montebello Islands, an archipelago in the Indian Ocean about eighty kilometers from Australia's western coast. A small armada including the floating laboratory HMS *Zeebrugge* and the hospital ship HMS *Tracker* set sail posthaste. The bomb was transported aboard an old frigate, the HMS *Plym*. The official purpose of the test was to simulate a nuclear attack on a British port. The bomb would be detonated aboard an enemy vessel, with the *Plym* playing that role. Once the fleet of ships had arrived at its destination, the *Plym* and its deadly cargo were anchored off Trimouille Island. A British propaganda film, *Operation Hurricane*, produced by the Ministry of Supply, documented the logistical complexity of the operation. In it, soldiers schlepped display-window mannequins dressed in uniform while others set up grenade launchers that would fire probes into the mushroom cloud to take measurements. Spooky music, written by star composer John Addison, played in the background. The film went into great detail about safety precautions. On the ships during the trip over from England, experts instructed the soldiers on how to deal with radioactive substances. At the site, doctors handed out protective clothing: thick rubber gloves and boots, along with gas masks. Many of the Royal Marines were issued unwieldy Geiger counters.

In image upon image, handsome, well-tanned soldiers, naked from the waist up, are shown waiting for the massive explosion. Taken from a low angle, these shots are eerily reminiscent of Leni Riefenstahl's notorious Nazi propaganda film *Olympia*. Even the scientists charged with arming and setting off the bomb strip their shirts off from one scene to the next.

Operation Hurricane would have us believe that the physicist who completed the last switching circuit before detonation was half-naked. Perhaps the idea was to suggest manly decisiveness—who knows? What is more important is that apparently most of the young men who took part in the test were not given any of the protective clothing and equipment depicted elsewhere in the film. Even lead vests, of course, would hardly have helped against the gamma rays emitted by the explosion—but, still, the wind could easily have shifted and blown radioactive particles onto the men's skin. As the shock wave moved across the Montebello lagoon, hundreds of sparsely clothed sailors sat on the decks of their ships only kilometers away, with their backs to the explosion. *Operation Hurricane* clearly shows them watching the giant atomic cloud from close proximity. Afterward, navy helicopters descended upon the contaminated area and took water samples. It is highly questionable whether the pilots were wearing protective suits. Tellingly, they are never on camera in the film.

Even today, there's little consensus as to whether the navy men's health was intentionally or unintentionally endangered. The British Ministry of Defense has been cleared of liability, but it was forced to defend itself for years in a class-action lawsuit brought by ex-servicemen who had participated in British nuclear tests in the South Pacific in the 1950s. The veterans claimed that they had been intentionally exposed to radiation—used as human guinea pigs. A British court initially ruled in favor of the ex-soldiers, but that judgment was overturned on appeal. In March 2012, the Supreme Court of the United Kingdom put an end to the matter by the closest of margins, four judges to three, ruling that the statute of limitations had expired, so that the ex-servicemen could not sue the Ministry of Defense.

Over the course of these trials, a controversial procedure became known to the public, creating a host of negative headlines. At the center of the controversy were the tests G1 and G2, which were part of "Operation Mosaic" in the Montebello Islands four years after Hurricane. Several Royal Marines testified that after the Mosaic detonation their ship had been ordered to sail directly into the radioactive cloud. One of the navy men, radio operator Bob Malcolmson, who was eighteen at the time, told the BBC that his ship, the HMS *Diana*, had its deck contaminated. Malcolmson claimed that men were sent below to get buckets and mops and then told to swab the deck with hot, soapy water. Several of those men later fell ill. "Several chaps lost teeth, and others lost their hair," Malcolmson testified. "So a lot of wives and sweethearts waited in Devonport to welcome back bald fiancés and bald boyfriends with a few teeth missing." In 1974, Malcolmson himself was diagnosed with polycythemia—a rare form of blood cancer he said his doctors linked with radiation exposure. By the start of the suit in 2008, many of Malcolmson's fellow navy men, including the captain of his ship, had died.

Rosenblatt Solicitors, the London law firm that represented 700 British veterans, presented a document from the British general staff that seemed to prove that the *Diana* was intentionally directed into the fallout zone. Detailed orders issued in 1953 stipulated that the precise effects of radiation on men and machinery were to be tested. Lorna Arnold, however, dismisses the document as non-incriminating, claiming that it referred to dummy ships that were supposed to be moored at Ground Zero, not to the *Diana*. Arnold writes:

> *Diana*'s movements were flexibly controlled and she intercepted the fallout, though not without some

problems of tracking. The prediction aspects of the experiment were judged to be satisfactory. The intended degree of fallout on the upper deck was achieved, and decontamination procedures were rapid and thorough. No member of the ship's company received a measurable exposure to gamma radiation.

But in 2008, the *Daily Telegraph* reported that of the *Diana's* crew of 240 sailors, only sixty were still alive, and of those who had died, more than a hundred had had cancer.

Other British nuclear experiments also caused controversy. In 1953, Britain began carrying out test detonations in the middle of the Australian mainland. The first test site, Emu Field, was located in a waterless desert in Southern Australia, even though water is crucial for decontaminating both men and equipment. Moreover, the absence of water wasn't the site's only drawback. The soft, sandy ground was completely ill-suited for the tests, since atomic explosions contaminate the sand and blow it around wildly. When British military leaders decided on the Emu Field location, they apparently failed to consider the possibility of people getting caught in radioactive sandstorms.

The Australian desert may have seemed uninhabited at first glance, but that was a fallacy. Ahead of the first atomic tests there, British scouts found indications that the area was in fact used by nomadic and semi-nomadic tribes. Attempts to steer these people clear of radioactive fallout were pathetic. For years, the welfare of Aborigines depended upon one man, Walter MacDougall. In 1947, this clever and popular fellow was charged with keeping Aborigines out of the Woomera rocket test site 480 kilometers from Emu Field. The task required him to cover some 3,000 kilometers of outback. In

1953, his area of responsibility was dramatically expanded to include the nuclear test site for the upcoming Totem experiments. MacDougall was very diligent about his job as protector of Aborigines, but it was impossible for any one individual to cover such a huge stretch of territory. His reports give a clear account of how unpredictable Aboriginal tribes' movements were and how much uncertainty arose as a result. After one of his missions, MacDougall concluded:

> A few natives still travel between the ranges and Ooldea. I saw a woman at Shirley Well, which is situated on the Officer River between the Everard and the Musgrave Range, whom I saw some time ago at Ooldea. The route they follow I believe is considerably west of Talleringa and Dingo Claypan [i.e. the Emu test site]. I suspect there is considerable travel east of the Warburton Range but west of Talleringa. There are natives occupying Tomkinson Range, and also the Mann, Peterman and the Rawlinson Ranges. Also they travel between the Warburton and Musgrave Ranges.

Some of the tribes that MacDougall visited had little or no contact with the outside world, and they were very reluctant to give information about their way of life. Inhabitants of the Everard Range, for example, were unwilling to talk about sites that had secret ritual significance. "It was not possible," MacDougall wrote, "to determine the degree of importance as commonly under such circumstances they depend upon the inaccessability [sic] of the actual sacred areas and the natural camouflage of water supplies to prevent trespassing."

Yet despite the uncertainty, MacDougall recommended that the Emu Field tests go ahead as planned. Looking back on the project years later, he regretted his involvement, admitting that "we might as well declare war on them and make a job of it." Yet MacDougall can hardly be held responsibility for what the British did in South and Central Australia. His reports were very critical of military plans, and he threatened on more than one occasion to go to the press with what he knew. It wasn't his fault that he had been sent into the vast outback on an impossible mission. After helping the inhabitants of one camp to move south, he made one last gigantic round, alerting the heads of all the missionary stations to the impending tests. After that, British commanders gave the go-ahead for the first Totem detonations.

On the day of that test, an Aboriginal woman named Lallie Lennon was searching for opals in the desert with her children. She had a small portable radio with her. At 6:00 a.m., when Totem 1 was detonated, her family was already wide awake. Later she recalled: "While I was waiting I made bottles up for that morning and we waited for it to go up. Holding our ears. We thought it was going to be all—you know—that scared." Suddenly there was a sound like thunder in the distance, and the earth shook. After a few seconds it was all over. But the worst was actually just beginning for the Lennons.

On the Wallatinna Aboriginal reserve, five-year-old Yami Lester heard a loud bang, after which the reserve's inhabitants began debating what was going on. Sometime that morning, probably between ten and eleven o'clock, a strange cloud approached the camp. "It was coming from the south, black-like smoke," Lester recalled. "I was thinking it might be a dust storm, but it was quiet, just moving, as it looks like, through the trees and above that again, you know. It was just rolling and moving quietly."

The phenomenon caught Lallie Lennon unawares. "Just

as we were getting ready for breakfast," she remembered, "this smoke, you know, you could see it between the trees coming through, it went right through, over us. Could smell the gunpowder smell. And on the tent it had this gray blacky sort of dust." The uncanny cloud disappeared just as quickly as it had arrived, but shortly after the incident, Lennon came down with a fever, and two of her children began to suffer from headaches, diarrhea, and vomiting. Her son complained about a burning sensation in his eyes. Two weeks later, Lallie Lennon and her son developed skin conditions. Strange red scabs formed on the boy's chest, and his face swelled up.

The black cloud caused no shortage of consternation in Wallatinna. As it approached, the Aboriginal elders believed they saw a *mamu*, a zombie with owl's eyes and murderously sharp teeth. They waved their spears in the air in an attempt to drive the evil spirit away from the camp. Others dug frantically in the sand dunes. Totally bewildered, Yami Lester sought refuge in one such trench. There he waited for the cloud, and its bestial stench, to pass. Others who were unable to bury themselves reported that the interior of the cloud had been white and that people's shadows had looked strange. A short time later, most of the Pitjantjatjara tribe fell ill. Their eyes became infected, many of them suffered from diarrhea, and older people developed skin afflictions. The tribe's drinking water suddenly tasted funny. Five days later, some of the Aborigines began to vomit bile. Dogs trailed after the seriously ill, waiting for them to collapse and die.

We can no longer reconstruct exactly how many people in the Aboriginal camp died from the effects of the black cloud. Death is subject to strict taboos among the Pitjantjatjara. They are not to speak about the dead, and the Pitjantjatjara system of counting was also next-to-impossible for the British to understand at the time. The best estimates put the number of dead at between thirty and forty. The symptoms, which were described identically in a number of eyewitness testimonials, suggest acute, massive radiation poisoning. It is especially fatal for the old and the weak.

Yami Lester was lucky to survive, but he was scarred by what had happened. He went blind in one eye, and his vision in the other deteriorated. His parents had to have his siblings lead him around by a stick. Today, Lester is over sixty and almost totally blind. We have no way of knowing whether his ailment was caused by radiation at Wallatinna or by a pre-existing condition.

What is beyond doubt is that those responsible for the nuclear tests based their calculations about the safety of the experiments on false assumptions. When estimating the effects of the maximum expected exposure to radiation, they used the average British person as a reference point. But the British were completely unlike Aborigines both physically and culturally. The average Briton was far larger and clothed from head to toe, whereas Aborigines were small in stature and usually almost completely naked. They very rarely wore shoes. The radiation thus came into direct contact with their skin. Moreover, a dosage of radiation that a Western European might have been capable of absorbing was much more lethal to Aborigines—especially children. British scientists' biggest mistake was to overlook drastic differences in nutrition. The food available to a desert nomad in Australia was by no means comparable to that of an Englishman or -woman. Aborigines foraged, hunted, and ate animals that the British would have considered absolutely inedible—snakes and various lizards, for example. The bodies of these creatures were particularly prone to absorbing radiation.

After the second Totem test, two security officers were sent into the contaminated area to "destroy," as one of them later put it, rabbits and dingoes. But the operation did little to protect the inhabitants of the region—it merely underscored how ignorant the British military was about Aboriginal habits. Even the Aboriginal protector, MacDougall, who was in constant contact with the tribes, failed to enlighten the military leadership, probably because he didn't realize the potential significance of what he knew. Where the Aborigines were concerned, the Australian Atomic Weapons Test Safety Committee (AWTSC) was an abject failure—not surprisingly, since the organization consisted of physicists and meteorologists who knew nothing about the biochemical effects of radiation. Authorities lacked crucial information about the clothing, nutritional habits, general physical constitution, means of communication, and movement patterns of Aborigines, most of whom still lived as hunters and gatherers in the 1950s.

British disinterest in and ignorance of Aborigines was also reflected in the military's choice of its next test site in 1955. The decision to drop bombs in the vicinity of the Trans-Australian Railway was made because the British thought it was necessary to carry out further tests but had realized that the lack of water made the Emu Field site unsuitable. The new site was christened Maralinga. Ironically, in one of the Aboriginal languages, the word means "field of thunder," but it had never been used in conjunction with this particular stretch of territory. The British often availed themselves of Aboriginal words to lend a mythical sound to their tests, but they rarely had a clue what myths they were referring to. An age-old Aboriginal path crossed straight through the test site. There were also several holy places in and around Maralinga, and Aborigines used some parts of the site as hunting grounds.

On May 14, 1957, the Milpuddie family was discovered in the crater left behind by Marcoo 1, and what became known as the "pom-pom incident" caused a political uproar. The AWTSC, in particular, came in for sharp criticism for the utter ineffectiveness of its safety precautions. Those responsible did their best to cover up what had happened. The Milpuddies were told that they had witnessed a secret white man's ceremony that they must never discuss with other white men.

Soldiers involved with the Maralinga tests have accused both the British military and the AWTSC of intentionally making do with insufficient safety standards. A number of sources allege that scientists in the contaminated "forward area" wore protective suits and gas masks, while ordinary soldiers had to go about their duties in regular uniforms. Many of these accusations were later disproven—but not all of them. An ex-soldier named Peter Webb who was stationed in Maralinga claimed he had been used as a human guinea pig. Webb was part of the so-called "Indoctrinee Force," a group of foot soldiers tasked at the tests with advancing onto the site directly after the detonation and carrying out various menial jobs. The combined British-Australian unit checked, for instance, on the condition of tanks, airplanes, and howitzers placed at Ground Zero. Webb recalled standing in the crater left behind by the third Maralinga bomb only hours after it had been detonated. The extreme heat of the nuclear firestorm had turned the red sand into spheres of glass. Webb said he kicked one of the strange formations out of curiosity and discovered it was as hard as rock. One eyewitness, a then-nineteen-year-old Australian named James Hutton, reported that he was ordered, a mere hour after three detonations, to dig up scientific instruments buried near Ground Zero. In the process, he and the other members of his unit swallowed radioactive dust. After the job was done, they fried eggs on the insides of their spades by holding them over a campfire.

As the medical historian Sue Roff has found out, the British military was keenly interested in the effects of radiation on human beings. During one nuclear exercise, commanders ordered recruits to roll around in radioactive dust to test which types of clothing offered the best protection against radiation. An eyewitness named Terry Toon reported seeing several personnel carriers discharging soldiers only five kilometers from the Marcoo crater. To Toon's bewilderment, the men fell to the ground in the desert sand. "We didn't know what was going on," Toon recalled. "They were all crawling around in red bulldust. It was ridiculous." A far larger action code-named "Operation Lighthouse" foresaw a test in which thousands of soldiers would be posted in trenches in the immediate vicinity of a mushroom cloud. Fortunately, this idea never became reality.

Dead soldiers and thousands of innocent civilians from various parts of Australia were also autopsied to determine whether their organs bore any traces of radiation. In the 2010 Redfern Inquiry report, the result of an investigative committee ordered by the House of Commons, the authors found that these people's organs had been taken from their bodies entirely without the consent of their relatives.

Today, only a few thousand of the 22,000 Englishmen and Australians who took part in the tests are still alive. The *Independent* newspaper has calculated the average life expectancy of soldiers stationed in the Montebello Islands and at Emu Field and Maralinga to be 55.5 years. Hardly any of these men or their families have ever received any compensation. One of them was a veteran's widow named Beth Campbell. After lengthy legal proceedings, the court decided that her husband's death from cancer was related to the inhalation of toxic substances during the so-called "Kittens" tests.

The Kittens tests did not involve any detonations, yet such "minor trials," as they were known, would prove all the more devastating in the long term. Some of the most harmful were the "Vixen A" tests, which simulated various nuclear accidents. They involved scientists setting fire to plutonium with gasoline, baking it in an electric oven and mixing it with conventional explosives to form a dirty bomb. This was done despite the facts that even a few milligrams of plutonium are potentially fatal for human beings, and that dosages in the micrograms can cause cancer decades after first exposure.

In 1967, the British tried to clean up the radioactive waste they had made in Australia with an action named "Operation Brumby." Military specialists plowed up the surface of the test site and dumped heavily contaminated soil into twenty-two concrete-lined pits. But a Royal Commission convened in 1985 to examine Maralinga's nuclear past found that such decontamination measures were insufficient. In particular, the areas where plutonium had been set alight were still highly contaminated. In 1996, after a wave of public pressure, a second, more thorough decontamination was carried out. When the employees of the specialist firms hired by the British government arrived at the edge of the test site, they found a ghost town. The workshops, laboratories, offices, fuel depots, and warehouses of the British military settlement at Maralinga were all still there, abandoned, as were the houses and bars, the swimming pool, and the hairdressing salon. Alan Parkinson—the Australian government's representative for the Maralinga cleanup project—described what the decontaminators saw:

The place is a treasure trove for people wishing to

collect old things such as electronic valves similar to those which used to power our wirelesses ... Anybody wanting telegraph wire or copper cable would have a ball; cable was ordered in job lots of 100 miles (160 km) and it is all still there ... But scroungers or souvenir hunters should take care. On a visit to Emu, we picked up one of those old silver threepenny pieces. When we checked it with a Geiger counter we found it was slightly radioactive—it must have been lying on the ground at one of the Totem sites when the bomb was exploded, and was made radioactive by neutron activation.

Within the test site itself, the technicians found the natural environment more or less intact. Kangaroos, for instance, were hopping around in the bush. But appearances were deceiving. In front of and behind the security fences set up during Operation Brumby, specialists measured high levels of radiation, and as in Semipalatinsk, Kazakhstan, clumps of uranium and plutonium were found just lying around. The pits dug in the 1960s were in terrible condition. Most of their caps were cracked and half-open. No one could determine how much radioactive material may have escaped, although the amounts were probably small. After the "Roller Coaster" experiments in the United States, which resembled those of Vixen A, 90 percent of the plutonium simply dispersed into the four winds. A British report put the amount of plutonium in the pits at one to three kilos and in the rest of the Maralinga test site at five to six kilos, while the remainder, around eleven to fourteen kilos, was deemed to have been blown off over Western Australia.

After reaching an agreement with the Australian government and the Tjarutja Aborigines, the experts decided to contain the Maralinga radiation using a complicated in situ vitrification (ISV) procedure. This technique entails melting contaminated rocks and soil with a high-voltage electrical current and then allowing them to cool. Plutonium and other radioactive substances in the soil congeal into glass blocks, which can then be buried.

Parkinson, however, says that over time the ISV procedure was deemed too expensive. It also carried risks, as an accident on March 21, 1999, showed. There was a sudden, powerful explosion in pit 17. The cap bulged, and a geyser of glowing liquid glass spurted up. Fortunately, it didn't hit anyone. Shortly thereafter, technicians stopped using the ISV procedure, although Parkinson, who quit the project in anger, claimed that the accident was merely used as a pretense for the authorities to save money. From that point on, radioactive material was simply bulldozed into shallow pits and covered. These measures clearly fell far short of a permanent solution for storing the radioactive waste. Tjarutja tribesmen, suspecting that the Australian government was trying to put off fully decontaminating their lands, protested against the change in plans. But their complaints were in vain.

Maralinga became a seemingly never-ending story. In addition to the British ex-servicemen who waged their protracted legal battle against the government, other groups have also pursued lawsuits. In January 2010, some former Australian soldiers, represented by a team led by lawyer Tom Goudkamp, began exploring the possibilities for suing the British government for injuries allegedly resulting from the aftereffects of radiation exposure at Australian nuclear test sites. That lawsuit was completely undermined after the British Supreme Court's 2012 ruling. A group of about fifty Aborigines also blamed the desert tests for their long-term health problems. A woman named Maureen Williams, for example, was just a baby when a mushroom cloud ascended into the sky above her homeland and has suffered all her life from a skin condition that may very well have resulted from exposure to radiation. Cherie Blair, wife of former British prime minister Tony Blair, agreed to take the case, but it too fell afoul of the Supreme Court's statute-of-limitations ruling. Whatever the truth may have been, it is very difficult to prove connections between radiation exposure and long-term illnesses.

The situation remains ambiguous. On the one hand, the British military made enormous and in some cases perhaps even criminal mistakes. On the other hand, some effective safety measures were clearly in place during the nuclear tests, and not everyone who took part in them was exposed to health risks. In the 1980s, the Royal Commission dismissed dozens of allegations that participants in the tests had been inadequately informed about what was going on. Some veterans also claimed that they had been sent into contaminated areas without any protection, whereas in reality there was no radiation whatsoever in the places where they had served.

In the end, it may well be impossible to determine precisely who fell physically ill because of the Maralinga, Emu Field, and Montebello Islands tests—and for many people any certainty will come too late. There is certainly no way to do justice to those who suffered psychological injuries and decades of worries and fears. It would be sensible for the government in London to agree to a generous collective settlement and draw a line under this less-than-glorious chapter in British history. That might prompt other nations to take a similar approach—in particular Western Europe's other nuclear power, France. In the course of *Gerboise verte* (Operation Green Jerboa), many French soldiers were exposed to radiation in the

Sahara. The French government has offered a collective ten million euros in compensation—far too little to offset the absence of compensation to those who suffered from dozens of French nuclear tests in Algeria and the South Pacific. As of fall 2012, lawsuits between veterans of French testing missions and the French government were ongoing. The situation with the British tests in Australia is thus just one part of a much larger phenomenon that continues to affect countless people.

In any case, the nuclear question will continue to be the subject of conflict and controversy on the red continent—and not just because of the British nuclear tests during the 1950s. Australia possesses huge uranium reserves, approximately 31 percent of the world's supply, according to the latest estimates. As human hunger for energy grows, so will the value of nuclear fuel. But Australia has never been entirely comfortable about exploiting this particular natural resource. The damage to the environment done by the excavation of the Rum Jungle mine in the north of the country in the early 1950s remains unforgotten. The disposal of the radioactive waste created by the project was catastrophic. Even today, there's still a controversy about who should be held responsible for that disaster.

Despite such experiences, new uranium mines are still being opened in the west of the country. In the past few years, the uranium industry has nearly quadrupled its output from that part of the continent. In the coming thirty years, the Yeelirrie mine, which is owned by the Cameco company, will produce some 90,000 tons of uranium ore. The courts have been deluged by protests from Aborigines, who in some cases have been able to win their battles for land rights.

Along with the ecological problems that uranium mining brings with it all over the globe, there is another source of controversy in Australia. The country has no nuclear power industry of its own, so its uranium is mined solely for export. That provides the foreign buyers with a material that can potentially be used for military purposes. In October 2012, Australia opened negotiations aimed at selling uranium to India, reversing a policy of not marketing radioactive material to countries that have refused to sign the Nuclear Non-Proliferation Treaty and that possess the atomic bomb. Proponents of the deal argue that the uranium is earmarked for civilian purposes and that its use will be closely monitored. Critics object that the use of Australian nuclear fuel in India's reactors indirectly supports the latter's nuclear-weapons program by freeing up other stocks of uranium for military use. The uranium mining trade remains a problematic undertaking. For countries like Australia, Canada, and Russia, it can be a source of immeasurable wealth. But the price is impossible to calculate. It is truly a Faustian bargain.

CHAPTER TEN

The Deadly Detours of Nuclear Medicine

On September 29, 1987, a stranger appeared in the regional Ministry of Health in the Brazilian province Goiás and shocked governmental officials there with an announcement that parts of the city Goiania, with its 1.3 million inhabitants, needed to be evacuated. He insisted on speaking personally with the minister of health. At first ministry employees tried to brush the man off, thinking that he must be deranged, but they gave in when he became more insistent and began to warn of an imminent catastrophe. At 2:30 p.m., the stranger, who turned out to be a highly qualified physicist, was brought to the minister of health, who immediately recognized the gravity of the situation. Three hours later, the head of Brazil's nuclear emergency task force NEC boarded a plane in Rio de Janeiro. At a stop in São Paulo, two further specialists, carrying radiation measuring devices, joined him on the aircraft. Meanwhile, the Goiás Ministry of Health contacted the tropical medicine clinic in Goiania, where several severely ill people were already being treated for what doctors believed was a mysterious allergy. The physicians were told that their patients were in fact suffering from acute radiation poisoning and should be immediately quarantined. By the end of day, the government had requisitioned Goiania's Olympic Stadium so that hundreds of other radiation-contaminated people could be examined and treated. The NEC team arrived in the city shortly after midnight and headed to an abandoned private medical clinic. The police subsequently evacuated several parts of Goiania, including neighborhoods surrounding three junkyards in some of the city's poorest districts. The organs of the Brazilian state worked quickly and efficiently, as the final act of one of the worst tragedies of the Atomic Age began to unfold.

The accident, which would eventually cost four people their lives, had begun in unspectacular fashion two years previously. In late 1985, the private radiology clinic Instituto Goiano de Radioterapia moved to a new location within the city. Because of a legal dispute, a piece of equipment—an Italian-built Cesapan F-3000 teletherapy unit used to administer radiation treatment to cancer patients—was left behind at the old site. Clinic proprietors had warned the relevant governmental authorities about the potential danger the device posed, but nothing was done. Part of the clinic was torn down, and the section of the building with the treatment rooms began to crumble. Homeless people began seeking shelter there. Over the course of 1987, a salvager of metal named Roberto dos Santos Alves heard rumors that there was plenty of valuable scrap at the site. On September 10, he and an assistant entered the building and discovered the perfectly intact Cesapan F-3000. Not suspecting any danger, and certainly not expecting radioactive cesium chloride, the two men disassembled the device over the next couple of days, unwittingly switching the Cesapan's radiation head to the "treatment" position and causing it to emit steady bursts of radiation. The two men then transported what amounted to a rapid-fire radiation cannon in a wheelbarrow to Alves's house and placed it under a mango tree. Shortly thereafter, both fell ill, assuming their sickness was related to something they had eaten. The assistant went to see a doctor, who diagnosed his diarrhea, vomiting, and swollen right hand as an allergic reaction. Meanwhile, Alves attacked the mysterious device with a screwdriver and punctured the capsule, about the size of a billiard ball, that contained the cesium. Mistaking the radioactive material within for gunpowder, he initially tried to light it on fire.

On September 18, Alves loaded up his wheelbarrow with parts from the Cesapan F-3000 and took them to Devair Alves Ferreira, the owner of a scrap-metal yard. When the latter entered his workshop that evening, he saw a strange bluish glow, which made him think that the material could be particularly valuable, and maybe even possess magic powers. He took it home for his family to admire, and in the days that followed, other relatives and friends came by to get a look at the mysterious glowing substance. Ferreira gave a portion of it to his brother and others close to him, and some of them rubbed it on their skin like Carnival glitter. Ferreira's wife fell ill and went to a doctor, complaining of the same symptoms as Alves and his assistant. She too was diagnosed with a food allergy and sent back home with the instructions to get some rest. Meanwhile, the symptoms suffered by Alves's assistant, Wagner Mota Pereira, had gotten much worse. He was taken to a hospital and then transferred to the Clinic for Tropical Medicine. Others patients with the same symptoms were soon to follow.

Ferreira's wife Gabriela, whose condition was deteriorating day by day, was the first one to blame the mysterious powder for the illness. She persuaded one of her husband's employees to gather up the remaining parts of the strange device and take them to a local hospital. The employee slung them in a plastic bag over his shoulder, suffering serious burns as a result. At the hospital, Gabriela Ferreira told the doctor on duty that a machine was killing her family. The doctor thought the radiation cannon was an X-ray machine and removed the plastic bag from his office, placing it on a chair near an adjacent wall. He then decided to consult with a physicist he knew who happened to be passing through Goiania. The next morning the physicist set off with a highly sensitive radiation detector, a scintillometer, to examine the suspicious scrap metal. The needle went off the scale before he had even arrived at the hospital. Thinking the device was defective, he proceeded to the local office of Nuclebrás, a public institution that monitors uranium mining in Brazil, and borrowed a second radiation detector. It too went wild before he reached the hospital. Realizing that something must be gravely amiss, the physicist convinced hospital officials to close the building. The fire department was called, and it took all of the physicist's persuasive power to convince them not to just throw the scrap metal in the river. He then succeeded in retracing the Cesapan F-3000's path back to the now totally contaminated scrapyard. Early that afternoon, the physicist paid a call to the Health Minister, and evacuations and other emergency measures were ordered.

Several houses, including the one belonging to Devair Alves Ferreira, had to be torn down. Tons of soil were dug up and taken away for disposal. The cesium, the plastic bag, and the chair were sealed in concrete and removed under the tightest security. In the city's Olympic Stadium, more than 200 people with symptoms of radiation poisoning showed up to be decontaminated and treated. Ten critically ill patients were flown from the Clinic for Tropical Medicine to Rio de Janeiro for better treatment. But for some, help came too late. Gabriela Ferreira succumbed to acute radiation poisoning, and two workers at the scrapyard also died. Perhaps most tragically,

Ferreira's brother had given his six-year-old daughter Leide several grams of the glowing substance to play with while she was sitting on the ground eating. She, too, died of the effects of being exposed to radioactive cesium. The final toll of the Goiania disaster: four dead and dozens ill and injured. But without the quick action taken by the physicist and local authorities, this catastrophe would have been far worse.

Irresponsible handling of nuclear medical technology could also lead to an accident in the United States or Europe, if probably not on the scale of the one in Goiania. The biggest danger is not radioactive material left lying around, but rather a bizarre inheritance from the 1970s and '80s. During these decades, scientists sent atoms on a very special trip through the human body, using plutonium to power pacemakers. This was a logical idea. The batteries in the first generation of pacemakers weren't very durable and had to be replaced every two years. Researchers thought they might be able to adapt technology, developed by space agencies, that harnessed the thermal energy created when plutonium decays in order to generate electricity. Atomic batteries of the sort also used in nautical buoys and remote radio stations can be made in minute sizes. By the 1970s, scientists felt they had perfected the technology, and it didn't require a great imaginative leap to picture such batteries powering pacemakers.

Plutonium 238, the element used in the batteries, has a half-life of 87.7 years—an atomic pacemaker thus lasted for a whole human lifetime without needing to be serviced or replaced. The radiation emitted did not harm the recipient, since plutonium gives off alpha particles. Whereas gamma rays can penetrate concrete walls, a piece of paper is enough to shield people from alpha particles, and in an atomic pacemaker, the poisonous radiation is contained within several sealed metal casings. It was highly unlikely that the plutonium would leak out and poison the patient. The situation was more problematic, however, when the patient eventually died. In our modern, mobile world, it is next-to-impossible to keep track of people. In 2009, for instance, an elderly Russian woman suddenly turned up at doctor's office in the eastern German city of Halle, where she struggled to communicate in her broken German that she was wearing an atomic pacemaker, which had been implanted by Soviet doctors. Where the woman is now, and whether she's even still alive, is unknown. Like other wearers of atomic pacemakers, perhaps she has died, and the plutonium-powered device is buried in a cemetery somewhere. Or the woman may have been cremated, in which case the pacemaker could have ended up in a garbage dump or in a home. This plutonium problem could have been solved had authorities kept tabs on where the people with atomic pacemakers lived and traveled. But neither the United States nor the Soviet Union—the two countries where the devices were implanted—was particularly diligent about documentation. Many of the pacemaker recipients, and with them the 200-milligram plutonium batteries, simply disappeared. No one knows where the lost pacemakers' journeys ended.

Nonetheless, even if an obvious mistake was made in this respect, the pacemakers were an undisputed blessing to their users. The problem was simply that people failed to think through all the consequences. Despite individual accidents and shortcomings, the use of radioactivity in medicine has clearly been a great success. X-ray machines, radioactive contrast agents, and radiation therapy have revolutionized the diagnosis and treatment of many diseases.

Still, there were detours along the path of progress that took doctors to their ethical limits and even beyond. Radioactive substances were not always employed for morally justifiable purposes. At the beginning of the Atomic Age, some researchers felt it was acceptable to inject test subjects, including a four-year-old child, with plutonium. By the mid-1940s, the United States was producing industrial amounts of weapons-grade atomic material. Thousands of people were involved in the various phases of this process at the nuclear factory at the Hanford Site in the state of Washington or at the Los Alamos nuclear-weapons facility. Those in charge were concerned about their workers. Little was known about the effects of radiation, and scientists had not yet determined which exposure levels were tolerable and which represented health risks. At first, researchers carried out experiments on animals, but they soon discovered that the results didn't necessarily apply to human beings. Atomic medical experts were interested above all in how much plutonium the body could eliminate after the substance had been ingested or absorbed. How good were our biological self-cleansing mechanisms? What percent of inhaled or ingested plutonium would be deposited within the body?

As Pulitzer Prize–winning journalist Eileen Welsome detailed in her chilling book *The Plutonium Files*, scientists in the United States tried to answer such questions by experimenting on thousands of human subjects. The researchers involved, some from prestigious institutions like the University of California, knew that there was no way they could attract volunteers for tests so obviously dangerous. Doctors, however, were not subject at the time to the strict code of behavior they are subject to today. There was no rule stipulating that physicians explain treatments to their patients. The researchers selected subjects who were undergoing treatment in hospitals for serious conditions and who were usually, though not always, considered terminally ill. Under the pretense of treating the patients, researchers injected them with poisonous, radioactive substances—sometimes in minute quantities, sometimes in larger doses. The patients had no idea they were being used as guinea pigs in a large-scale experiment. At regular intervals, the researchers took blood, stool, and urine samples and measured the amount of plutonium that the patients' bodies had eliminated. One of the eighteen subjects in the first series of experiments was a four-year-old Australian boy named Simeon Shaw, who suffered from leukemia. His parents had him flown to the United States so that he could get the latest and best treatment, but instead of trying to save his life, doctors injected him with a cocktail of plutonium 239, yttrium and cerium. Shortly after returning to Australia, he died an excruciating death.

The cynical calculations underlying this top-secret project, however, had a flaw. Some of the subjects who the researchers thought were terminally ill survived their diseases and were eventually discharged, unaware that their bodies now contained radioactive substances. One subject—a twenty-four-year-old woman named Mary Jeanne Connell—was used as a guinea pig even though she was nowhere close to dying. She had checked herself into Strong Memorial Hospital at the University of Rochester in New York for anemia and low weight. There she got a nasty surprise. A nurse showed her some laboratory rats, mice, and rabbits in cages and asked her if she wanted to "help humanity." Thinking it was a joke, Connell laughed and declined. Nonetheless, when she woke up on the morning of October 1, 1946, she found herself strapped to a gurney, surrounded by doctors. One of them held a needle full of orange fluid that seemed to make the physicians skittish. This was injected into Connell's bloodstream. The ampule contained uranium salts, exposing Connell to a level of radiation fifty-seven times as great as the average person would absorb in a lifetime. Years later, she remembered that her body felt like it was lying on hot coals for a couple of days. She nearly lost consciousness. She was kept in the clinic for weeks. A woman watched over her at her bedside day and night.

By 1997, Connell was the only survivor among twelve people known to have been subjects of involuntary experiments at Strong Memorial. The United States government's human radiation experiments are rightly considered the most shameful chapter in the history of American medicine. As many as 20,000 people may have been subjected in some form to involuntary tests, some of which were carried out long after World War II. In some cases, radioactive substances were injected. In others, the test subjects were made to swallow or inhale radiation. In still others, the test subject's entire body was bombarded with radiation from a large machine concealed within the wooden paneling of a room.

The methods used were similar to those employed by the secret police in Communist East Germany at their prison in the town of Gera. After the end of the Cold War, a concealed X-ray machine was discovered in a room in which inmates were photographed. The machine was pointed straight at the backs of the prisoners' heads. Several former East German dissidents have speculated that they were intentionally exposed to radiation as a form of punishment.

Prison inmates in the United States were also bombarded with radiation, as were children in a home for the mentally handicapped. Another group affected were the Inupiat Eskimos in Alaska. In 1950, near the Brooks Range, once the proposed site of Project Chariot and an underground nuclear test, the Inupiat were given pills full of iodine 131 under the pretense of explaining to them how their bodies worked. In fact, the researchers wanted to learn more about the effects of the radioactive substance. In 1996, by way of compensation, the Inupiats were promised a free medical examination, but this was surely insufficient to restore their faith in the government and its medical system.

CHAPTER ELEVEN

Broken Arrows

Mars Bluff, a small South Carolina village snugly surrounded by pine woods, is just about the most peaceful place imaginable. If you take the main road eastward, at some point you'll come upon the hamlets Pee Dee and Ketchuptown. About fifteen miles farther on is Florence, the nearest small city in the region. This is typical backwoods America, sleepy and seemingly endless. During the afternoon of March 11, 1958, a rail conductor named Walter Gregg was piddling around in his yard, while his nine-year-old daughters Helen Elisabeth and Francis Mabel played with their cousin Ella Davis. At approximately 4:30 p.m., Gregg heard the roar of three B-47 fighter jets. That was nothing unusual. There was a big Air Force base across the Georgia border, and pilots often flew sorties over the sparsely populated areas in inland South Carolina. Gregg watched the planes for a bit and went into his garage. Just then there was a massive explosion. For a minute, the conductor thought the sky was falling. Smoke was everywhere. When the fumes and the stench had cleared, Gregg saw a scene of total devastation. One wall of his garage was missing, the roof of his house had been completely blown off, and a car that a salesman happened to be driving by had been picked up and turned around 180 degrees. There was a crater, twenty-two meters wide, in Gregg's backward. And in the middle of it was a partially destroyed Mark 36 atomic bomb.

This was the same model that had been dropped on Nagasaki, although the Mars Bluff warhead was more modern, powerful, and dangerous. When the tear-shaped bomb had hit the ground in Gregg's backyard, its chemical trigger had detonated, hurling 300 pounds of high explosives through the air with incredible force. Gregg's first thoughts when he came to his senses were of his daughters. Miraculously, both were unharmed, but their cousin Ella was bleeding from a cut to the forehead. Just to be on the safe side, the family took her to the hospital. Gregg, himself a U.S. Air Force veteran, knew that the explosion had been caused by a bomb. He only found out later, when decontamination teams arrived at his property, that the bomb had been a nuclear warhead accidentally dropped from one of the B-47s.

The spot where Gregg's backyard once was is open to visitors, and the crater has been preserved. To get there, you take U.S. highway 327 west out to Crater Road, pass a trailer park, and then turn onto a tiny path leading through the underbrush. It's full of dark, brackish water. That has nothing to do with the bomb. In the past, foresters used the crater as a pit for burning tree trunks. There used to be signs here directing tourists to the unusual attraction, but they've all been stolen.

No amount of imagination is sufficient to see this muddy little pond as a memorial warning future generations of a nuclear near-catastrophe. The danger that the bomb presented was restricted to the chemical detonator and several slightly radioactive components that could only spread contamination in their immediate vicinity. The Mark 36 atom bomb belonged to a generation of warheads whose plutonium core was kept in a separate capsule called the "birdcage." To arm the bomb, the crew of the warplane had to take the capsule from the birdcage and install it—this was known as an "open pit." Flight crews had explicit standing orders that this was only to be done in wartime. In the case of Mars Bluff, the bomb fell to earth without its most deadly component, so there was no chance of a nuclear chain reaction. The safety system passed the test.

As political tensions increased during the Cold War and the reaction times to a presumed nuclear first strike by the enemy became shorter and shorter, the superpowers began to rethink bomb designs. Later models of atomic bombs are so-called "sealed-pit weapons" with the nuclear core integrated into the bomb itself. The warheads were easier to use and quicker to activate, so both the United States and the Soviet Union needed more sophisticated safety systems. In an attempt to prevent accidental detonations, engineers developed "intelligent" bombs that would recognize their environments. If the parameters were wrong, the switching circuits needed for detonation would be interrupted. Warheads designed to be carried by ICBMs had to measure extreme velocities, changes in air pressure, and a descent back into the atmosphere before they were allowed to arm themselves. Bombs designed for airplanes were equipped with comparable sensors. Speed and air-pressure detectors effectively ruled out the chance of a nuclear bomb accidentally going off in ground storage. Yet despite all the precautionary measures, there was at least one well-documented instance in which the world narrowly avoided an inadvertent detonation of a nuclear weapon.

On January 24, 1961, three B-52 bombers took off from a U.S. Air Force base. Officially the flights were for training, but in reality they were part of the top-secret Chrome Dome mission. Starting in 1960, the Pentagon had planes carrying atomic weapons in the air at all times to counter any attack by the Soviet Union. Around the clock, as though in a relay race, bombers took off and landed. One of the B-52s, which was carrying two Mark 39 hydrogen bombs, reported an oil leak in the motor of its left landing flap, but otherwise there were no irregularities. The planes flew in two large circles over the United States and the Atlantic, repeatedly refueling in midair. The third time, a filler neck broke off of the plane with the oil leak and kerosene began pouring from its right wing. Within a minute and a half, the aircraft's tank number 4 was completely empty, and the amount of fuel in its main tank had sunk by 25 percent. The plane's equilibrium was disrupted so that its flight became unstable. The U.S. Air Force base in Wayne County, North Carolina, gave the plane emergency permission to land, but the pilot was instructed to circle around and dump more kerosene to reduce the risk of fire. As the plane was descending toward the base, its landing gear down, there was a bang, and the bomber nosedived to its right. Five of the eight crewmembers escaped using ejector seats, with all but one of them surviving. Seconds later, the bomber broke apart. The back end just missed a house belonging to farmer Bud Tyndall and careened into a field in Eureka, near the town of Goldsboro, setting it on fire. Tyndall later said he thought the whole world was burning.

The hydrogen bombs had uncoupled themselves from their carrier systems while the plane was still in the air. A parachute opened to slow the fall of one warhead; the second plowed twenty feet into the muddy ground of the field going 700 miles per hour. At first, the U.S. Air Force denied that there had been significant risk of an accidental detonation. But in 1983, former Secretary of Defense Robert McNamara admitted that the bomb with the parachute had gone through all but one of the seven automatic steps toward detonation. A declassified chart supports this statement, although it details three of four switches that had been thrown. One was a circuit that closes after changes in atmospheric pressure. The triggering of the parachute was also part of the firing sequence, since Mark 39 bombs were designed to explode in the air over their targets. A clock that would count down the seconds until the explosion had also been activated.

Today, the 1961 Goldsboro Crash is still regarded as the most serious accident ever caused by the U.S. military, but unfortunately it is hardly the only incident of its type. There were dozens of accidents in which nuclear bombs were damaged or even lost during the Cold War—in both the West and the East. In military parlance, these were known as "broken arrows."

In the aftermath of the Goldsboro Crash, cleanup teams never succeeded in locating more than fragments of the so-called "primary," or first stage, of the bomb that had buried itself underground. This fission-based explosive device was nothing other than an ordinary atomic bomb—the nuclear trigger that set off the fusion explosion of a hydrogen bomb, in which hydrogen is transformed to helium just like in the sun. It is the enormous heat and radiation produced by the primary that allows fusion to arise in the secondary. And this second, larger part of the Goldsboro warhead was completely missing. Experts gave up in frustration after digging down more than fifteen meters. No unusual radiation was ever detected at the site, so they simply buried the impact crater under concrete. Heavy metal cylinders, or remnants of such, must still be somewhere down there. Today, all you can see are fields of beans revealing no trace of the accident. The U.S. military purchased an easement over the field. Eureka farmers are still allowed to plant there, but digging any deeper than necessary is strictly prohibited. The Army Corps of Engineers continues to test the water for elevated radiation levels, but none has ever been detected. This broken arrow looks as though it will continue to have no permanent consequences.

Other people caught up in accidents weren't as lucky as the residents of Goldsboro, though, even though the security failures weren't as grave. On January 17, 1966, a B-52 with four nuclear warheads on board was out on a Chrome Dome mission in the skies above the Mediterranean. The southern patrol route was the hardest slog of this mission, with pilots flying from the eastern coast of the United States all the way to Albania. There were six such flights every day to the southernmost edge of the Warsaw Pact countries. Because the routes were so long, the B-52s had to be repeatedly refueled in midair. At 10:30 a.m., during one such refueling, the bomber collided with the KC-135 refueling plane. The KC-135 exploded, killing everyone on board, while four crewmembers from the B-52 escaped using ejector seats and were saved by hastily summoned Spanish fishing boats. Radar navigator Ivens Buchanan landed hard on solid ground and had to be taken to a Spanish hospital. All four of the hydrogen bombs detached from the disintegrating B-52 before it crashed, but they did not activate. Three of them opened their parachutes and landed near the village of Palomares. One of those three was later found more or less undamaged in a river, while the conventional explosives in the others detonated. The power of the explosions blew several kilos of plutonium in all directions. At least 56.8 acres were contaminated, although recent studies indicate the affected area was actually quite a bit larger. Even today, authorities continue to monitor Palomares's inhabitants. There are Geiger counters next to their tomato fields.

The fourth bomb caused huge problems. Eventually it emerged that its parachute, too, had opened, and that the bomb had been blown out over the sea. Through happy co-incidence, a fisherman named Francisco Simó Orts saw it sink into the water—the Spanish press later nicknamed him "Bomber Paco." As troops moved in to decontaminate the ground at Palomares, a fleet of nineteen U.S. naval vessels was dispatched to the area. Nonetheless, it took eighty days of in-tense searching for the Navy men to locate the nuclear bomb. It had come to rest on the side of an underwater cliff 800 meters below the surface. The first attempt to raise it from the cragged terrain failed. The tension during the second attempt was so great that the commander of the diving team fainted.

Even after the bomb had been raised, the Palomares bro-ken arrow had not been completely blunted. The measures needed to combat radioactive contamination continue to cost large sums of money. Radiation can still be measured in the area, despite the removal of more than 1,400 tons of soil. There is no consensus about whether cancer rates in the region have been elevated. Optimists like to cite the example of an old woman, at whose feet one of the bombs virtually fell, and who achieved fame in the tabloid press as "Spain's most radioactive grandmother." The *señora* died at the stately age of ninety-two—of natural causes.

Over the course of the 1960s, people began to view the Chrome Dome missions more critically. The reason wasn't the Palomares accident, but rather another dangerous incident that happened near a remote U.S. Air Force base in Thule, Greenland. The base is located at the foot of a glacier between a pair of rivers. Denmark provided this area as an exclave to the United States, which built a radar station there in 1960. The station monitored airspace over the Soviet Union so as to alert the Pentagon should the Kremlin launch a first strike. Thule was the heart of the American early-warning system, and as part of Operation Chrome Dome, B-52s armed with thermonuclear warheads circled above the base at all times. They were deployed as a deterrent. Had the Soviets launched a conventional or nuclear attack against the radar station, the bombers that were part of the "Thule Monitor" mission would report that fact to Washington and then set course for the Soviet Union to retaliate. The Danish public knew nothing of this. Officially, Thule was a nuclear-free area. The government in Copenhagen, however, silently accepted the presence of "special weapons" there.

As perilous as the mission of the B-52 crews may have sounded, their everyday routines were tedious. The endless circling flights over Thule wore airmen down, as did the long Arctic winters and frigid temperatures. The G-model B-52 was heated with the exhaust from its engines. But because of a design flaw, it was as hot as a sauna on the lower deck where the navigators sat, while the deck above it remained cold and clammy. On January 21, 1968, the men on the upper deck were particularly frigid, and one of them asked the pilot for permission to turn the heat up to emergency—the maximum level. Permission was granted, and the heating system was set on full. On the lower deck, burning hot air issued from the metal ventilator, which was mounted under one of the seats. Even before the system was turned up, it was too hot for the navigator sitting there, and he had stuffed his jacket under the seat. After a few minutes, the nylon in this winter parka caught fire.

The navigator panicked. While he did succeed in warning the flight deck, he was unable to bring the rapidly spreading fire under control. When the kerosene in one of the fuel tanks ignited and the cockpit filled with smoke, the pilot had no choice but to issue the bail-out order. Because the crew on this flight numbered seven instead of the usual six, there was one ejector seat too few. The copilot, who had been resting on a cot, had to jump from the plane using an ordinary parachute. The harness of the parachute caught fire, the belt burned through, and the copilot plunged to his death. The other crewmembers landed more or less unharmed in the freezing Greenland wilderness. The B-52 remained aloft for a while, then banked steeply north of Thule and crashed into the ice with its payload of four Mark 28 F1 hydrogen bombs.

The weather being clear, eyewitnesses could see the crash from a distance. One of them was a young Danish logistics expert, Jens Zinglersen, who was a member of the paramilitary group "Royal Greenland Trade Department." This troop of 1,000 men was responsible for keeping Thule supplied. On January 21, Zinglersen had just returned home from a long shift at the base and was looking from the peninsula where Thule was located out at an uninhabited island. The visibility was good despite the perpetual twilight of the polar winter. Zinglersen was a nature enthusiast who loved the wilds of Greenland, so it wasn't unusual for him to be standing outside admiring the raw beauty of the land. What he suddenly saw in the sky, however, was anything but usual.

Years later, I interviewed Zinglersen, and he gave me an account of what happened. All at once, he said, there was a trail of fire in the ice in front of Appat Island, then a clap of thunder that shook him to the soles of his feet. "Good God," he thought. Zinglersen immediately recognized that a plane had crashed—what else could have caused the fire and noise? He called the base commander and reported that a "broken arrow" had occurred. Zinglersen had always suspected that there were nuclear weapons in Thule, although he'd never had the courage to ask directly. At the time he knew nothing about the effects of radioactivity, and staring at the crash site, he didn't think about potential dangers. If someone gets in trouble in the Arctic, people help—it's that simple. Zinglersen set off with two teams of sled dogs. Four Eskimos accompanied him.

The would-be rescuers plotted a zigzag course across the ice, searching for survivors. Zinglersen took the lead so as to be able to warn his fellow searchers about any perils. Later, ten more dogsleds arrived from an Inuit settlement. But all they found was a 275-meter-long black spot on the ice that smelled of kerosene. Wreckage was scattered around, but most of the pieces were scarcely larger than a pack of cigarettes. Zinglersen and the others could also see where there had been a hole in the ice, but the hole had frozen over. There were parachutes at various spots. Since Zinglersen didn't know that hydrogen bombs contain parachutes, he assumed that they had belonged to the crew of the plane. Mistakenly, he concluded that no one had survived. The crew members who had ejected from the plane were discovered later. One of them had a severe case of frostbite.

Ten hours later, when Zinglersen returned to Thule, the Strategic Air Command from Omaha, Nebraska had taken control of operations. The Danish outdoorsman, clad in thick animal skins, was taken directly to see two-star general Richard Hunziker, a handsome man whose face reminded Zinglersen of a Roman statue. When he reported on what he had seen, Hunziker snapped, "I don't believe a word of it." Zinglersen was furious. He hadn't had anything to eat for hours and was completely exhausted from the search. He didn't owe any Americans any explanations. He turned around and was about to storm out the door, when a uniformed medical officer blocked his way, asking "Can I quickly measure your boots?" The officer held a device, most likely a Geiger counter, down to Zinglersen's boots. Then he turned to the general and said: "He's been out there." Hunziker's attitude changed completely. He addressed Zinglersen politely and even stuck a thick cigar in Zinglersen's mouth. Later, Hunziker took Zinglersen to six nuclear technicians and had him lead them to the accident site. In addition, with the help of the Inuit, Zinglersen constructed a helicopter landing pad at the site and several igloos where guards could escape the cold.

Only after he had returned from the ice the second time was Zinglersen taken to a decontamination station. There was a bit of a quarrel when the Inuit refused to take off their contaminated furs. For them, as hunters, warm clothing was key to survival. Only when the Americans promised that the furs would simply be cleaned and then returned did the Inuit agree to hand them over. In fact, the furs were transported by plane to the United States and were never seen again. The Inuit were told that the long hairs of the furs made them impossible to clean. Zinglersen replaced them with new polar-bear furs and sent the Americans the bill. Since Zinglersen could speak Kalaallisut, he was asked to warn the Inuit against going near the crash site. He was supposed to tell them that there were dangerous munitions in the area. Zinglersen choose another word instead. He knew that the Inuit would understand munitions as bullets, and that they would head to the site by the dozens, if they thought there were free bullets lying on the ground.

The ice that Zinglersen and his Inuit helpers had spent hours traversing was heavily contaminated by the plutonium from the bombs as well as from other radioactive components. In the coming weeks, Danes who worked for the Trade Department, wearing no protective breathing apparatus, would take away more than two million liters of radioactive ice. It was then taken by ship for safekeeping in the United States. The cancer rate among Trade Department workers is 50 percent higher than in the Danish population at large, and over the years, Zinglersen and the others repeatedly sued for access to information and free health monitoring. In 2008, their case went all the way to the European Court of Justice. But Zinglersen and the others lost. There was a catch. A control group of Danes who also worked in Thule but who had no longer been there for the cleanup operation also showed an elevated rate of cancer. Expert consultants concluded that the cancer proclivity was the result of the hard working conditions and the men's above-average alcohol and cigarette consumption. An association formed by former Danish cleanup workers at Thule rejects this finding. In any case, the Danish government has never offered the men a compensation package, including information and free medical checkups, of the sort that was agreed upon in Palomares.

In addition to the debate about possible long-term health consequences, Thule also gave birth to two other controversies. The revelation that there were indeed nuclear weapons at Thule was a major embarrassment for the Danish government. At the time of the accident, authorities tried to tell the public that the B-52 had been on a mission outside Danish territory when it ran into trouble and asked for emergency permission to land in Thule. It wasn't until the 1990s that the whole incident emerged from secrecy. People in Denmark still speak today of "Thulegate."

The second controversy was what became of the hydrogen bomb with the serial number 78252, which was probably mounted on the bottom left of the bomb bay. On November 10, 2008, citing recently declassified military sources, the BBC reported that the bomb had melted through the ice and had sunk to the bottom of the sea. The headline "Mystery of lost U.S. nuclear bomb" suggested that the warhead was lying intact in the waters just off Thule. In the article itself, however, the author admitted that what had been lost was not a functional bomb, but the fusion secondary. The primary had been destroyed by the conventional explosion unleashed by the crash. As sources, the author named 300 documents that the Pentagon had declassified, in censored form, during the 1980s and '90s. These papers detail the sequence of events leading to the accident and the cleanup operations, code-named "Operation Crested Ice," that took place in the months that followed.

Several passages in these documents indicate that bomb components had gone missing. An early report filed in late January 1968 refers to the hole in the ice at the accident scene that, as Zinglersen described, had tracks from a parachute at its edges. The anonymous author of the report speculates that a primary or secondary might have melted its way through the ice. According to the BBC, the U.S. military commenced an extended search in an effort to find the missing bomb parts. A submarine was sent to comb the waters near and under the crash site—without success. The Pentagon issued strict instructions that for the purpose of communication with the Danish government, these activities were to be referred to only as a survey. In parts of the BBC report, the author dramatized the situation. There is no evidence in the Pentagon documents, for instance, that the leaders of the operation became panicky as the weather turned bad and the search had to be suspended. One of the eyewitnesses cited by the BBC took a rather stoic view of an unsuccessful dive. It was disappointing, the man said, that not all of the bomb parts had been found. But he also reasoned that if his own people couldn't find the military wreckage, it would be just as difficult for others to do so.

The BBC report didn't contain much new information—
Denmark's flagship newspaper *Jyllands-Posten* had run a story
about the "missing bomb" eight years before, using exactly the
same sources. But the British report made waves. The Danish
government felt compelled to commission an expert report on
the incident from the Danish Institute of International Study.
The title of that report, authored by historian Svend Aage
Christensen, made no secret of what it would conclude: "The
Marshal's Baton—There is no bomb. There was no bomb.
They were not looking for a bomb." Christensen decided that
the part must have been what he calls a "marshal's baton," a
hollow fifty-centimeter-long uranium cylinder whose function, most people think, was to start the fusion of heavy water in the second, thermonuclear stage of a hydrogen bomb,
much as a spark plug triggers an engine. (Much of the information concerning the construction of nuclear weapons is still
kept top secret.) Christensen concluded that it was most likely
this, and not a complete secondary or an intact bomb, that the
Americans were searching for near Thule.

Christensen also refuted several points from the BBC report:

1) According to Pentagon documents, the bomb with
the serial number 78252 would have been in extremely
poor condition after the conventional explosives inside the primary detonated during the crash. Most
likely, the radioactive material in the secondary had
simply been pulverized. For that reason alone it was
highly improbable that there would be a large, intact
part of the bomb on the ocean floor.

2) Despite the fact that the explosions caused by the

conventional warhead and thousands of liters of kerosene had destroyed most of the plane and its deadly cargo, search teams had located radioactive material with a weight of roughly 94 percent of what one would expect from three secondaries. That did not mean that the fourth one was missing, but rather that it had probably been destroyed. The parts that had been discovered could not be precisely assigned to any specific bomb. Some of them may have come from the fourth bomb, number 78252.

3) The amount of plutonium found by Crested Ice was roughly what one would expect from four bombs of this type.

4) A STAR III submarine had been deployed in the search, even though it had one weak grabber capable of lifting only twenty kilos at a time. That would have made it an unsuitable choice for a mission intended to recover a secondary fusion core presumably shrouded in a thick layer of uranium.

5) Nowhere did the Pentagon documents refer to a missing bomb, only to a missing part, and many of those documents were given the low-level security classification "not for foreign nationals."

Christensen did not deny that the U.S. military had been searching for something, but he concluded that it must have been a small object. The historian surmised that it could have been a component from bomb number 453171, whose secondary had been discovered intact. With this bomb, in contrast to the utterly demolished 78252, cleanup crews would have noticed if a part was missing.

Christensen's arguments are persuasive. The tenth operational status report from September 10, 1968, leaves no doubt that the search teams were looking for a relatively lightweight object: "It was further speculated that the missing [word or words censored], in view of its ballistic characteristics, may have come to rest beyond the observed concentration of the heavy debris." This description simply doesn't fit a squat secondary most probably inside a massive uranium shell.

Nonetheless, Christensen could only deal in probabilities, and his report remained speculative at a number of crucial junctures. The idea that the plutonium found is roughly the amount contained in four bombs is based on a brief marginal notation. How much plutonium a hydrogen bomb actually contains is top-secret information. The total amount (7.5 kilos) of plutonium that the report calculates as being about right for four bombs seems too small and would assume that the bomb's fuel was surrounded by neutron reflectors. Otherwise, such a small amount of plutonium could hardly have achieved critical mass. It's not generally known whether the secondary of a hydrogen bomb also contains plutonium. It is conceivable that there is a plutonium rod within a hollow uranium detonator. If the thermonuclear stage does *not* contain plutonium, however, Christensen's calculations are irrelevant to the questions of whether a secondary went missing.

Moreover, while it makes sense that a STAR III submarine would not have been suitable for the task of recovering a secondary, that observation is hardly the last word in the story. Different equipment is often used for the two parts of a search-and-recovery operation, and there are a number of potential reasons why the Pentagon documents make no mention of recovery equipment. Perhaps the documents concerning this topic were never declassified. Or maybe the U.S. military feared arousing Danish distrust by sending in massive recovery machinery. Or perhaps the search-and-recovery team simply lacked the necessary resources. The year 1968 was an *annus horribilis* for the Pentagon. It began with North Korea capturing a U.S. spy ship, and a short time later the Tet offensive commenced in Vietnam. In the spring, a nuclear submarine sank in the Atlantic. It was the worst possible point in time to send a search-and-recovery fleet like the one deployed at Palomares to Greenland.

Jens Zinglersen, who spent ten years in Thule and who has also studied the Pentagon documents, remains skeptical. He objects that the most important passages in the declassified material have been blacked out and accuses the American military of lying. He recently wrote an account of his own rebutting the DIIS report. It is doubtful that we will ever know the whole truth about what happened. In 2003, Danish experts from the Risø National Laboratory at the Technical University of Denmark searched the floor of the ocean off Thule with a modern sonar device. The only unusual thing that it turned up was a five-meter-square object. The object was not identified, but it certainly did not match the description of any bomb part. Otherwise there was nothing out of the ordinary.

No matter how much uncertainty may surround the Thule broken arrow, at least there are some documents on which theories can be based. That's thanks to the U.S. Freedom of Information Act. Other countries have no such legal provisions. There's very little known about Soviet broken arrows, to say nothing of those from the world's smaller nuclear powers. We can safely assume that the real number of accidents and mishaps is far greater than the number of those that have become public. Along with incidents in which nuclear bombs have been damaged or destroyed, there are cases in which warheads have gone missing entirely. Some thirty to forty atomic and hydrogen bombs have officially "disappeared"—though some experts speculate that the number may be as high as fifty. Such figures are, of course, highly speculative. But if we consider that a single Soviet Typhoon submarine was armed with up to twenty ICBMs, and each of them was equipped with several warheads, the minimum numbers no longer seem so improbable.

Rumors persist, for instance, that a Soviet submarine laid twenty nuclear mines on the ocean floor in the Bay of Naples. As bizarre as this story may seem, there are other well-documented accounts of atomic weaponry simply disappearing into the sea. One of them was the USS *Scorpion*, the American attack submarine that sank in 1968. On board were two nuclear Mark 45 Astor torpedoes. The U.S. Navy assures us that the wreck of the *Scorpion* poses no danger, despite the sub's plutonium cargo. A Soviet Golf II class submarine, the *K-129*, also sank in the Pacific in 1968. It was carrying several nuclear missiles to be fired at the West Coast of the United States in the event of an all-out war. Using a network of sonar devices in the ocean, the United States managed to locate the wreckage. Instead of passing the information along to the Soviets, one of the most expensive CIA missions of all time was called into being. The aim of "Project Azorian" was to raise the *K-129* from the ocean floor to analyze codebooks and its onboard nuclear warheads, missiles, and systems.

Doubts abounded early on during the project. Security experts objected not only to the horrendous costs, but to the risk of heightened political tensions should the secret mission be uncovered. Nonetheless, Azorian had the support of both

Richard Nixon and Henry Kissinger. The navy built a recovery ship, the *Glomar Explorer*, with a special, giant claw to grab hold of the submarine and lift it up to the ocean's surface. With the suspicious Soviets keeping a close eye on the Pacific, the American military thought up a complex cover story. Officially, the *Glomar Explorer* was an experimental vessel, testing the possibilities for mining manganese nodules from the ocean floor. It remained anyone's guess why sane people would try to extract the nickel-and-iron nodules, which are more easily accessible elsewhere, from a depth of five kilometers, and the CIA knew that the implausibility of the endeavor would raise Communist mistrust. So they recruited a front man for the project who was known to be no stranger to outlandish ideas: Howard Hughes. The billionaire was a notorious eccentric who was constantly funding megalomaniac schemes. The *Glomar Explorer* thus putatively put out to sea in the service of the Hughes Tool Company.

Unforeseen problems cropped up right from the get-go. While docked in the Chilean port of Valparaíso, the crew got caught in the middle of the Pinochet coup d'etat, and the U.S. government had to intervene before the vessel was allowed to sail on. When it finally arrived at the spot where the submarine had gone down, the crew's work was repeatedly interrupted. Once they had to take a fellow sailor on board who had suffered a heart attack on another ship. Then Soviet navy boats arrived, and a Russian helicopter began circling overhead. Even today, despite a raft of CIA documents being declassified in 2010, we don't know precisely how successful the project was. Crucial passages from the documents have been blacked out. Journalists Sherry Sontag and Christopher Drew, who researched Project Azorian, concluded that the submarine broke in half during the recovery operation, but that the crew of the *Glomar Explorer* had succeeded in raising its bow, which contained two nuclear torpedoes. There was also video evidence, presented to Russian President Boris Yeltsin in the 1990s, that the project was not a complete failure. The video showed several Soviet sailors who had been raised from the *K-129* being given a proper burial at sea. If we believe this version of events, it would mean that there are still four torpedoes with nuclear warheads lying on the ocean floor. They, too, are a legacy of the Cold War.

Many of these self-created problems will continue to occupy humanity for thousands of years to come. And the nuclear follies described in this book are only the tip of the iceberg. No one knows how many nasty surprises from the past are currently waiting for us on land or under the sea. This much is certain: The legacy of the Cold War remains a giant liability for the world's health.

History will show whether the "hot" war on Hiroshima and Nagasaki was the terrible high point or a terrifying foretaste of mankind's misuse of nuclear technology. Unfortunately, there is little evidence that people have gotten much wiser or more responsible since 1945. Nuclear technology continues to spread. In the not-too-distant future, Iran may develop a nuclear bomb of its own, setting off an atomic arms race in the Middle East involving Saudi Arabia and other countries. Israel already has a significant nuclear arsenal, and it doesn't require much imagination to envision a "hot" war breaking out in that region. Tensions could also escalate between the nuclear powers India and Pakistan or between North and South Korea. To make things worse, several Asian nations are now embroiled in a race to develop long-range ballistic missiles. Meanwhile, the technology of uranium enrichment recently took a quantum leap forward. Isotope separation by laser excitation may soon replace centrifuges and will make nuclear arms programs more efficient and potentially easier to conceal. The new START agreement between the United States and Russia notwithstanding, we are still far from a world free of nuclear weapons. Global relations of power are changing at a furious pace, creating new risks. At some point, nation-states may no longer have a monopoly on nuclear weapons, and organizations of various stripes—in the worst case, terrorist groups—could acquire bombs. At the same time, despite the Fukushima disaster, the world's dwindling oil reserves and global warming could lead to a renaissance of nuclear power. If nuclear is preferred over renewable energy sources, greater and greater areas will be destroyed by uranium mines, the problem of how to store nuclear waste will intensify, the danger of disruptions will increase, and larger amounts of weapons-grade plutonium produced by nuclear reactors will be created. It is anybody's guess which new chapters of atomic idiocy will be added to a tale already full of folly.

Notes

CHAPTER ONE AFTER THE BOMB, THE WORLD'S MOST DANGEROUS INVENTION

8 "The country that succeeded in building..."; Eichler, p. 2

9 "When asked by Armaments Minister Albert Speer..."; http://www.brookings.edu/projects/archive/nucweapons/manhattan.aspx

10 "In Riehl's view, insufficient state..."; Riehl, p. 19

10 "J. Robert Oppenheimer, the father of the American bomb..."; Bird/Sherwin, p. 267

10 "The controversial German historian Rainer Karlsch..."; Karlsch, p. 54

11 "Trees were reportedly uprooted..."; Ibid., p. 56

11 "The German governmental organization that examined..."; http://www.ptb.de/de/aktuelles/archiv/presseinfos/pi2006/pitext/pi060215.htm

11 "T Force soldiers abducted any research scientists..."; *The Guardian*, August 29, 2007

13 "This operation, directed by..."; Trutanow, p. 72

13 "Riehl later recalled..."; Riehl, p. 9

14 "But the German scientists would agree..."; Zippe, p. 73

15 "Years later, at an official state reception..."; http://www.bbc.co.uk/radio4/science/zippetype.shtml

15 "Anything further the researchers... "; Riehl, p. 20

16 "The more probable explanation..."; Zippe, p. 75

17 "Nothing at all moved me . . ."; Steenbeck, p. 163 f.

19 "The only disturbances. . ."; Zippe, p. 79

19–20 "On account of my. . ."; Ibid., p. 96

20 "Zippe claimed that he. . ."; Ibid., p. 138

22 "It was only down to a huge amount of luck. . ."; Ibid., p. 99

22 "The Germans working on other parts. . ."; Steenbeck, p. 272

24 "But the incident, together with. . .";
http://www.fas.org/irp/cia/product/zippe.pdf

27 "No one noticed that Khan. . ."; Koch, pp. 48–50

27 "In addition, Hennie, who like her husband . . .";
http://www.bbc.co.uk/radio4/science/zippetype.shtml

27 "The CES Kalthof company in Freiburg. . ."; Koch, p. 75

29 "With a kitchen knife you can peel. . .";
http://www.bbc.co.uk/radio4/science/zippetype.shtml

CHAPTER TWO THE RED BOMB

31 "At the same time, the surrounding countryside. . ."; As reported by the
Los Angeles Times on November 27, 1990

34 "Two others had already died of the same causes. . ."; Cited in Trutanow, p. 170 f.

34 "Sometime later, Zhukov went out to inspect Ground Zero. . ."; *Der
Spiegel*, September 19, 1994, p. 164

35 "However long the Soviets may have used. . ."; Bradley, p. 12

35 "Shocked, Anatoly P. Sklyanin. . ."; http://articles.latimes.com/1990-11-
27/news/wr-5585_1_nuclear-waste-radioactive-wastes-nuclear-program
Accessed on August 26, 2012

36 "A 1992 report drawn up by the American Department of Energy. . .";
Bradley, p. 12

36 "The DoE lists the following radioactive remnants. . ."; http://www.
osti.gov/bridge/servlets/purl/6821704-E6i0Wd/6821704.pdf. Accessed
on August 26, 2012

38–39 "The stress of…"; http://www-ns.iaea.org/downloads/rw/waste-safety/north-test-site-final.pdf Accessed on August 26, 2012

39 "In the 2004 IAEA report…"; Ibid., p. 99

39 "When called upon by radio to specify their intentions…"; *Die Zeit*, October 19, 1990

41 "We were their guinea pigs…"; http://english.aljazera.net/programmes/witness/2009/08/20098273257926312.html

41 "A lot of people were injured by shards of glass…"; Trutanow, p. 24

42 "No one told the villagers that the nuclear tests…"; http://english.aljazeera.net/programmes/witness/2009/08/20098273257926312.html

43 "That same year, Czech police pulled…"; Jones/McDonough, p. 28

44 "It had Kazakh laborers covering particularly…" *Science*, May 23, 2003, pp. 1220–1224

44 "Threat Reduction program for destroying weapons…"; http://www.wikileaks.ch/cable/2010/02/10ASTANA251.html#par32

46 "Unmanned drones, sent by the United States Defense Department…"; *New York Times*, May 21, 2011

CHAPTER THREE THE MYTH OF TACTICAL NUCLEAR WAR

48 "As part of tests in the earlier 'Hardtack' series…"; http://www.archive.org/details/OperationHARDTACK_HighAltitudeTests1958

48 "One of them, a naval officer…"; Hoerlin, p. 13 f.

49 "As a result, the 'Uracca' experiment…"; Ibid., p. 46

50 "Over time, the radioactivity released into the upper reaches…"; Roedel, p. 171 f.

53 "ADMs have thus been rightly described…"; Holdstock/Barnaby, p. 39

54 "With those numbers of dead…"; Strachan, p. 195

54 "In the 1990s, a former member of the Soviet General Staff, Adrian Danilevich…"; http://www.gwu.edu/~nsarchiv/nukevault/ebb285/vol%20i1%20Danilevich.pdf, p. 31

56 "For those reasons, the authors of the UN study concluded…"; Potter/Sokov/Müller/Schaper, p. 4

56 "While the Cold War did see the creation of atomic mines. . ."; The Atomic Testing Archive in Nevada has now made declassified films of the tests available to the general public. See http://www.nv.doe.gov/library/films/testfilms.aspx

57 "Whether or not such weapons. . ."; Smith, p. 435

57 "Nonetheless, in order to destroy an underground bunker. . ."; http://www.spiegel.de/wissenschaft/mensch/0,1518,353994,00.html

CHAPTER FOUR THE RADIOACTIVE COWBOY, OR HOW ALASKA GOT THE BOMB

59 "He then allegedly dragged her. . ."; Holston, p. 108

60 "Allegedly, the veteran gunslinger. . ."; Roberts/Olson, p. 410

62 "Meanwhile, Wayne's wife and Hayward. . ."; Roberts/Olson, p. 411

63 "Cancer killed Dick Powell. . ."; People, November 10, 1980

64 "Normally, in a group of 220 people. . ."; People, November 10, 1980

64 "There is absolutely no doubt. . ."; Brown/Broeske, p. 261

65 "If the gigantic nuclear explosion is detonated. . ."; Cited in Schumacher, p. 108

65 "Alerted to the dangers of nuclear testing. . ."; Kohlhoff, p. 73

67 "In a U.S. Department of Energy memo from 1950. . ."; AEC Memo "Additional Test Site" of December 13, 1950, p. 3, https://www.osti.gov/opennet/servlets/purl/16388781-LywrwX/16388781.pdf

68 "That idea was eventually abandoned. . ."; Der Spiegel, June 21, 1996

68 "The bombs that were supposed. . ."; Kohlhoff, pp. 36–38

69 "Recently released documents do show. . ."; The Guardian, May 24, 2010

70 "According to scientific estimates, there were. . ."; Kohlhoff, p. 59

71 "Hours after the explosion. . ."; Ibid., p. 88

72–73 "The seal provided by the island. . ."; Johnson/Stewart, pp. 12–28

CHAPTER FIVE SWORDS INTO PLOWSHARES

84 "The solution they came up with was to build. . ."; Teller, et. al., p. vi.

84 "The study concluded that...", Ibid., p. 226

85 "As the cost of transporting material to the moon...", Ibid., p. 285

85 "And Teller was not alone...", For U.S. efforts, see http://www.guardian.co.uk/science/2000/may/14/spaceexploration.theobserver; for Soviet efforts, see http://www.iol.co.za/news/world/russia-wanted-nuclear-bomb-on-moon-1.4078#.UL2162dHUmQ

86 "Only the communists' discipline...", Teller (2001), p. 15

86 "He had fled there via Copenhagen...", Ibid., 182 f.

87 "Nuclear physicist Robert Serber termed him..."; Bird/Sherwin, p. 279

87 "Although I began explaining all those reasons to Bethe..."; Teller (2001), p. 177

89 "Those who favored dropping the bomb...", Ibid., 2007

91 "He also voiced the hope that...", *Day at Night*, May 8, 1974, http://www.youtube.com/watch?v=z8uZKsoPv68

92 "The costs of constructing the port...", O'Neill, p. 40

92 "Their audience was disconcerted...", http://arcticcircle.uconn.edu/VirtualClassroom/Chariot/vandegraft.html

93 "Moreover, there was one other minor detail...", Kirsch, p. 61

93 "Even a detonation with half the planned strength of Chariot..."; Ibid. p. 48

94 "They didn't reveal that the traditional Inupiat...", Chance, p. 5

94 "That communication made it clear ...", Kirsch, p. 87.

97 "John N. Wolfe, the head of the environmental section..."; http://www.hss.doe.gov/healthsafety/ihs/marshall/collection/data/ihp1b/7712_.pdf, p. 34 f.

97 "The third quarterly AEC report...", http://www.hss.doe.gov/healthsafety/ihs/marshall/collection/data/ihp1b/7714_.pdf , p. 89 f.

98 "The press, too, began to take an...", Kirsch, p. 92

100 "Another village resident...", O'Neill, p. 133

101 "The scientists had no official permission...", Chance, p. 9

101 "On October 17, 1992, the following press...", Ibid. p. 16

103 "A representative of the Alaskan government. . ." http://www.hss.doe. gov/healthsafety/ohre/roadmap/events/stakeholders/20.html, p. 1

103 "The road, of course, had to be. . ."; Kirsch, p. 113

104 "As predicted. . ."; Ibid., p. 126

104 "The crater was large enough. . ."; https://www.osti.gov/opennet/serv- lets/purl/16009208-gxpM3b/16009208.pdf, p. 2

104 "As a 1983 Department of Energy dossier revealed. . ."; https://www. osti.gov/opennet/detail.jsp?osti_id=16009208&query_id=1

106 "Hence routes normally considered. . ." Teller, et. al., p. 239 f

106 "But after a year at the latest. . ."; Kirsch, p. 174

108 "And twenty-one bombs were detonated. . ."; All figures from Carlisle, "Soviet PNES"

109 "The film commentator explains that. . ."; http://www.youtube. com/ watch?v=T97xMNtITTY

109 "No one knows what happened. . ."; http://www.fas.org/sgp/othergov/ doe/lanl/osti/408695.pdf, p. 8

109 "The water in the lake contains high levels of tritium. . ."; Salbu/Skip- perud, p. 51

110 "The Russians, who are not as limited. . ."; Teller (1981), p. 173

111 "Summing up the positives of his plan, Bassler wrote. . ."; Bassler, p. 6

111 "By the end of the 1970s. . ."; http://www.wasserbau. tu-darmstadt.de/ wasserbau/ueberuns/geschichte_1/index. de.jsp

CHAPTER SIX THE DOOMSDAY MACHINE

113 "The radioactive dust would reach California in about a day. . ."; The *New York Times*, April 7, 1954

116 "What is the practical importance of this?"; Transcript, University of Chicago and NBC: "The University of Chicago Round Table—The Facts about the Hydrogen Bomb," part two of the series "How We Can Make Peace," February 26, 1950, p. 7 f.

117 "Human beings, the aging physicist predicted. . ."; Smith, p. 375

119 "Submarines could bring gigantic bombs of this type to strategic points..."; *Popular Science*, September 1962, p. 214 f.

119 "Edward Teller was incensed at reports like this..."; Smith, p. 391

119 "Once again, the spark was a major article..."; The *New York Times*, October 8, 1993

121 "Former members of Soviet missile divisions allegedly told him..."; Rosenbaum, p. 92 f.

122 "Moreover, the political and military leadership would have had to consult..."; *Wired*, September 21, 2009

123 "Yarynych speaks of 1984, while Katayev talks about the early 1980s..."; http://www.gwu.edu/~nsarchiv/nukevault/ebb285/vol%2011%20Kataev.pdf, p. 101

123 "They use a process of three person control..."; Cited in Maggelet/Oskins, Vol. 2, p. 208 f.

124 "At the other end of the scale, by the way, was Great Britain..."; http://news.bbc.co.uk/2/hi/programmes/newsnight/7097101.stm

124–125 "Dr. Surikov responded that..."; http://www.gwu.edu/~nsarchiv/nukevault/ebb285/vol%20II%20Surikov.pdf, p. 134 f.

CHAPTER SEVEN FLYING REACTORS

128 "One of the experts involved described the enterprise..."; http://www.loyola.edu/departments/academics/political-science/strategic-intelligence/intel/cosmos954.pdf

128–129 "*Time* magazine calculated that..."; *Time*, February 6, 1978

129–130 "They had arrived complete with all their gear in two C-141 military cargo aircraft..."; http://www.qbristow.ca/Life_Divided/chap7a.html#1

132 "In fact, they had only been exposed to a small amount of radiation..."; *Time*, February 6, 1978

133 "All of them were highly radioactive..."; May, p. 279

134 "The atomic oven was basically..."; *Der Spiegel*, August 25, 1954

134 "It could fly at its top speed all the time..."; *Time*, June 5, 1948

136 "The radiation intensities encountered in nuclear reactors..."; *Time*, June 5, 1948

CHAPTER EIGHT HOW SAFE IS SAFE?

153 "Gillon also tirelessly lobbied for another major project..."; Wrong, p. 140

155 "When journalist Michela Wrong interviewed Kalenga..."; *T he Financial Times*, August 21, 1999

155 "Reassuringly, the uranium contained by the rod is nowhere near sufficient..."; Daly/Parachini/Rosenau, pp. 65–67

155 "And while it's difficult to convert the sort of material used in research reactors into fuel for bombs..."; http://nnsa.energy.gov/sites/default/files/nnsa/inlinefiles/2011_NNSA_Strat_Plan.pdf, p. 4

156 "But it's nearly impossible to determine..."; http://www.bbc.co.uk/2/hi/africa/6430031.stm

156 "The visibly nervous technical director of the facility..."; http://rnw24.nl/africa/bulletin/dr-congos-lone-nuclear-reactor-idle-safe-expert

157 "Today, the radioactive battery is probably somewhere..."; May, p. 193 f.

159 "That is why national programs like the one recently announced by Saudi Arabia..."; Cooke, p. 173

160–161 "Whereas plumbers can easily check the flame of a gas boiler..."; Cooke, p. 359

161 "In 1992, BBC journalist Adam Curtis made an investigative documentary film..."; "A for Atom" from the 1992 BBC series "Pandora's Box," http://www.bbc.co.uk/blogs/adamcurtis/2011/03/a_is_for_atom.html

163 "In 1975, for example, there was a disruption..."; May, p. 267

163 "The Iranian interior minister denied that this had taken place..."; *The Daily Telegraph*, January 16, 2011

CHAPTER NINE ATOMIC AUSTRALIA

167 "Her second child died of a brain tumor. . ."; Holdstock/Barnaby, p. 87

167 "The fact that the Aborigines' interests. . ."; Arnold, pp. 278–279

168 "The British decision to make an atomic bomb. . ."; Gowing,
Vol. 1, p. 185

169 "Their estimate was wrong. . ."; See Medawar/Pyke, p. 220

169 "On the contrary, one year after the end of World War II. . .";
Arnold, p. 6

170 "A British propaganda film, *Operation Hurricane*. . ."; The Ministry of
Supply was the military's procurement branch.

170 "In it, soldiers schlepped display-window mannequins. . ."; Addison
would later have a successful career in Hollywood, composing the
music for, among other films, *A Bridge Too Far*.

171 "As the shock wave moved across the Montebello lagoon. . .";
http://www.nationalarchives.gov.uk/films/1951to1964/
filmpage_oper_hurr.htm

171 "In March 2012, the Supreme Court of the United Kingdom. . .";
http://www.guardian.co.uk/law/2012/mar/14/pacific-atomic-test-
survivors-mod

172 "Several chaps lost teeth, and others lost their hair. . .";
http://news.bbc.co.uk/2/hi/uk_news/7273738.stm

173 "But in 2008, the *Daily Telegraph* reported. . .";
http://www.telegraph.co.uk/news/uknews/1574676/
HMS-Diana-the-ship-that-went-nuclear.html

173 "For years, the welfare of Aborigines depended. . .";
http://www.news.bbc.co.uk/2/hi/uk_news/7273738.stm

174 "A few natives still travel between. . ."; Arnold, p. 132

174 "It was not possible," MacDougall wrote. . ."; Rockets were also tested
independently of the nuclear weapons tests.

174–175 "Yet despite the uncertainty, MacDougall recommended. . ."; Cited in
Holdstock/Barnaby, p. 89

175   "While I was waiting I made bottles...";
      https://cooberpedyregionaltimes.wordpress.com/2010/04/28/
      maralinga-the-black-mist-incident/

175   "I was thinking it might be a dust storm..."; Royal Commission,
      Vol. 1, p. 174

176–177 "They are not to speak about the dead..."; Ibid., p. 175.

178   "Authorities lacked crucial information..."; Holdstock/Barnaby.

178–179 "There were also several holy places..."; Ibid., p. 84

179   "The Milpuddies were told that..."; See Cross/Hudson, p. 75

179–180 "After the job was done..."; *The Independent*, June 15, 2001

180   "We didn't know what was going on..."; http://www.reocities.com/
      jimgreen3/maralinga2.html

180   "In the 2010 Redfern Inquiry report..."; http://nuclearhistory.
      wordpress.com/2010/11/17/the-redfern-inquiry-into-human-tissue-
      analysis-in-uk-nuclear-facilities-volume-1-report-findings/

180   "The *Independent* newspaper has calculated..."; *The Independent*,
      June 15, 2001; The article does not name a source, and the informa-
      tion could not be verified.

182   "The place is a treasure trove for people..."; Parkinson, pp. 16–17.

182   "A British report put the amount..."; Ibid., p. 152

185   "Despite such experiences, new uranium mines..."; http://www.
      miningaustralia.com.au/news/wa-uraniumexploration-boom

186   "In October 2012, Australia opened negotiations..."; http://articles.
      economictimes.indiatimes.com/2012-10-17/news/34525690_1_
      uranium-sales-kazakhstan-and-canada-global-nuclear-trade

CHAPTER TEN THE DEADLY DETOURS
OF NUCLEAR MEDICINE

191   "But without the quick action..."; IAEA (1988), p. 31 f.

192   "In 2009, for instance, an elderly Russian woman..."; *Der Spiegel*,
      November 23, 2009

193 "As Pulitzer Prize–winning journalist Eileen Welsome. . ."; Welsome p. 7; Subcommittee on Energy Conservation and Power, p. 2 f.

194 "Shortly after returning to Australia. . ."; Welsome, pp. 153–155

195 "A woman watched over her. . ."; Ibid., p. 445 f.; see also http://www.people.com/people/archive/article/0,,20122017,00.html

195 "Several former East German dissidents. . ."; *Der Spiegel*, May 17, 1999

196 "In 1996, by way of compensation, the Inupiats. . ."; "Minutes of the Stakeholder's Workshop (February 27, 1996), http://www.hss.doe.gov/healthsafety/ohre/roadmap/events/stakeholders/20.html, p. 1 f.

CHAPTER ELEVEN BROKEN ARROWS

197 "There was a big Air Force base across. . ."; Miller, p. 317

198–199 "To arm the bomb, the crew. . ."; Maggelet/Oskins, Vol. 1, p. 5 f.

200 "Tyndall later said. . ."; Miller, p. 320

202 "There are Geiger counters. . ."; *Die Zeit*, January 16, 2011

203 "The tension during the second attempt. . ."; *Faceplate: The Official Newsletter for the Divers and Salvors of the US Navy*, Vol. 9, Nr. 2 (2006), p. 15

204 "After a few minutes, the nylon. . ."; Other accounts say it was a seat cushion stuffed into the heating grate that started the fire.

204 "When the kerosene in one of the fuel tanks. . ."; Maggelet/Oskins, Vol. 1, p. 230

205 "Years later, I interviewed Zinglersen. . ."; Interview of Zinglersen carried out by Rudolph Herzog on September 23, 2011, in the Harz mountains in Germany.

212 "It was further speculated. . ."; Cited in: Christensen, p. 94

213 "Otherwise there was nothing. . ."; Nielsen/Roos, p. 10

214 "Rumors persist, for instance. . ."; http://www.independent.co.uk/news/world/europe/soviet-navy-left-20-nuclear-warheads-in-bay-of-naples-6150280.html

214 "The U.S. Navy assures us. . ."; http://www.ibiblio.org/pub/academic/history/marshall/military/navy/USN_sub_losses.txt

Works Cited

Arnold, Lorna, *Britain, Australia and the Bomb: The Nuclear Tests and Their Aftermath*. Basingstoke, 2006 (Palgrave Macmillan)

Bassler, Friedrich, "Neue Vorschläge für die Entwicklung der Kattara-Senke." *Wasserbau-Mitteilungen*, Darmstadt, 1975, pp. 1–18

Bird, Kai and Sherwin, Martin J., *J. Robert Oppenheimer: Die Biographie*. Berlin, 2010 (Ullstein; original: *Prometheus: The Triumph and Tragedy of J. Robert Oppenheimer*)

Bradley, D.J., "Radioactive Contamination of the Arctic Region, Baltic Sea and the Sea of Japan from Activities in the former Soviet Union, Rickland, 1992," http://www.osti.gov/bridge/product.biblio.jsp?osti_id=6821704

Bristow, Quentin, "A Life Divided," Chapter 7: "Operation Morning Light. A personal account" (1995), http://www.qbristow.ca/Life_Divided/chap7a.html#1

Brown, Peter Harry and Broeske, Pat H., *Howard Hughes: The Untold Story*. Cambridge, MA, 1996 (Harvard University Press)

Carlisle, Rodney P., *Encyclopedia of the Atomic Age*. New York, 2001 (Facts on File)

Chance, Norman, "Project Chariot. The Nuclear Legacy of Cape Thompson, Alaska," http://arcticcircle.uconn.edu/VirtualClassroom/Chariot/chariot.html

Christensen, Svend Aage, *DIIS Report: The Marshall's Baton—There Is No Bomb. There Was No Bomb. They Were Not Looking for a Bomb.* Copenhagen, 2009

Cooke, Stephanie, *Atom. Die Geschichte des nuklearen Irrtums.* Cologne, 2011 (Kiepenheuer & Witsch; English original: *In Mortal Hands*)

Cross, Roger and Hudson, Avon, *Beyond Belief: The British Bomb Tests: Australia's Veterans Speak Out.* Kent Town, South Australia, 2005 (Wakefield Press)

Daly, Sara/Parachini, John/Rosenau, William, *Aum Shinrikyo, Al Qaeda and the Kinshasa Reactor. Implications of Three Case Studies for Combating Nuclear Terrorism.* (Study of the RAND Corporation) Santa Monica, 2005

Eichler, Jürgen, "Uranmaschinen und Atombombenpläne in Deutschland bis 1945. Physikalische und technische Aspekte, 2005," http://public.tfh-berlin.de/~hironaga/vorlesungs-scripts/entwicklung_a-bombe_in_dtld.pdf

Goodchild, Peter, *Edward Teller: The Real Dr. Strangelove.* London, 2004 (Weidenfeld & Nicolson)

Gowing, Margaret, *Independence and Deterrence: Britain and Atomic Energy.* London, 1974 (Macmillan)

Hoerlin, Herman, "United States High Altitude Test Experiences. A Review Emphasizing the Impact on the Environment," Los Alamos, 1976, http://www.fas.org/sgp/othergov/doe/lanl/docs1/00322994.pdf

Holdstock, Douglas and Barnaby, Frank, *The British Nuclear Weapons Programme 1952–2002.* London, 2003 (Frank Cass)

Holston, Kim R., *Susan Hayward, Her Films and Life.* Jefferson, North Carolina, 2002 (McFarland)

IAEA, "The Radiological Accident in Goiania" (Vienna, 1988), http://www-pub.iaea.org/MTCD/publications/PubDetAR.asp?pubId=3684

IAEA Division of Nuclear Safety and Security, "Nuclear Explosions in the USSR—The North Test Site, Reference Material," Vs. 4.

(Vienna, 2004), http://www-ns.iaea.org/downloads/rw/waste-safety/north-test-site-final.pdf

Johnson, Mark and Stewart, Colin, "Results from the Amchitka Oceanographic Study," Fairbanks, Alaska/Keyport, Wash., 2005, http://www.cresp.org/Amchitka/Amchitka_Final_Report/finalreport/05Append_bathymetry/5A_bathymetry_5_25_05.pdf

Jones, Rodney W. and McDonough, Mark G., *Tracking Nuclear Proliferation: A Guide in Maps and Charts.* Washington, D.C., 1998 (Carnegie Endowment for International Peace)

Karlsch, Rainer, "The German Atomic Projects in the Second World War," in *Nuclear Proliferation: History and Present Problems.* Berlin, 2009, pp. 51–61 (Max Planck Institut für Wissenschaftsgeschichte)

Kirsch, Scott, *Proving Grounds: Project Plowshare and the Unrealized Dream of Nuclear Earthmoving.* Piscataway, NJ, 2005 (Rutgers University Press)

Knoth, Robert and deJong, Antoinette, *Certificate no. 000358—Nuclear Devastation in Kazakhstan, Ukraine, Belarus, the Urals and Siberia.* Amsterdam, 2006 (Ram Distribution)

Koch, Egmont R., *Tödliche Pläne: Wie die Atombombe in die falschen Hände gelangte.* Berlin, 2007 (Aufbau)

Kohlhoff, Dean W., *Amchitka and the Bomb: Nuclear Testing in Alaska.* Seattle, 2002 (University of Washington Press)

Kramer, Fritz W., *Bikini: Atomares Testgebiet im Pazifik.* Berlin, 2000 (Wagenbach)

Lester, Yami, *Yami: The Autobiography of Yami Lester.* Alice Springs, 1993 (IAD Press)

Lindner, Konrad, *Ein Atomphysiker erzählt: Edward Teller zwischen Leipzig und Livermore.* Leipzig, 1998 (Universität Leipzig)

Light, Michael, *100 Suns and the Nuclear Sublime.* New York, 2003 (Knopf)

Maggelet, Michael H. and Oskins, James C., *Broken Arrow. Vol. I. The Declassified History of American Nuclear Weapons Accidents.* Milton Keynes, 2007 (lulu.com)

Maggelet, Michael H. and Oskins, James C., *Broken Arrow. Vol. II. A Disclosure of U.S., Soviet, and British Nuclear Weapon Incidents and Accidents, 1945–2008.* Milton Keynes, 2010 (lulu.com)

May, John, *Das Greenpeace–Handbuch des Atomzeitalters: Daten, Fakten, Katastrophen.* Munich, 1989 (Droemer Knaur)

Medawar, J.S. and Pyke, David, *Hitler's Gift: The True Story of the Scientists Expelled by the Nazi Regime.* New York, 2001 (Arcade Publishing)

Miller, Richard Lee, *Under the Cloud: The Decades of Nuclear Testing.* New York, 1991 (Two Sixty Press)

Nielsen, Sven P. and Roos, Per, *Thule 2003: Analysis of Radioactive Contamination.* Roskilde, 2006 (Risø National Laboratory)

O'Neill, Dan, *The Firecracker Boys: H-Bombs, Inupiat Eskimos, and the Roots of the Environmental Movement.* Philadelphia, 1994 (Basic Books)

Parkinson, Alan, *Maralinga: Australia's Nuclear Waste Cover-Up.* Sydney, 2007 (ABC Books)

Potter, William C./Sokov, Nikolai/Müller, Harald/Schapper, Annette, *Tactical nuclear weapons: options for control.* Monterey, 2000 (Monterey Institute of International Studies)

Randi, James, *An Encyclopedia of Claims, Frauds, Hoaxes of the Occult and Supernatural.* New York, 1997 (St. Martin's Griffin)

Riehl, Nikolaus, *Zehn Jahre im Goldenen Käfig: Erlebnisse beim Aufbau der sowjetischen Uran-Industrie.* Stuttgart, 1988 (Riederer)

Roberts, Randy and Olson, James Stuart, *John Wayne: American.* New York, 1995 (Free Press)

Roedel, Walter, *Physik unserer Umwelt: Die Atmosphäre.* Heidelberg, 2000 (Springer Berlin)

Royal Commission into the British Nuclear Tests in Australia, "Final Report" (1985), http://www.ret.gov.au/Search/retresults. aspx?k=royal%20commission%20report

Rosenbaum, Ron, *How the End Begins: The Road to a Nuclear World War III*. New York, 2011 (Simon & Schuster)

Salbu, Brit and Skipperud, Lindis (eds.), *Nuclear Risk in Central Asia*. Dordrecht, 2008 (Springer)

Schumacher, Geoff; *Sun, Sin and Suburbia: An Essential History of Modern Las Vegas*. Las Vegas, 2004 (Stephens Press LLC)

Shapiro, Charles S./Kiselev, Valerie I./Zaitsev, Eugene V., *Nuclear Tests: Long Term Consequences in the Semipalatinsk/Altai Region*. San Francisco, Bernal, 1994 (Tyrolia)

Smith, P.D., *Doomsday Men: The Real Dr. Strangelove and the Dream of the Superweapon*. London, 2007 (Penguin)

Sontag, Sherry and Drew, Christopher, *Blind Man's Bluff: The Untold Story of American Submarine Espionage*. New York, 1998 (HarperCollins)

Steenbeck, Max, *Impulse und Wirkungen: Schritte auf meinem Lebensweg*. Berlin, 1977 (Verlag der Nation)

Strachan, Hew, *European Armies and the Conduct of War*. Boston, 1988 (Routledge)

Subcommittee on Energy Conservation and Power, US House of Representatives, "American Nuclear Guinea Pigs: Three Decades of Radiation Experiments on US Citizens." (Washington D.C., 1996), www.hss.doe/healthsafety/ohre/roadmap/overview/07035013.html

Teller, Edward, *Memoirs: A Twentieth-Century Journey in Science and Politics*. Cambridge, 2001 (Perseus)

Teller, Edward, *Energie für ein neues Jahrtausend*. Berlin, 1981 (Ullstein; original: *Energy from Heaven and Earth*)

Teller, Edward/Talley, Wilson K./Higgins, Gary H./Johnson, Gerald W., *The Constructive Uses of Nuclear Explosives*. New York, 1968 (McGraw-Hill)

Trutanow, Igor, *Die Hölle von Semipalatinsk*. Berlin, 1992 (Aufbau)

University of Chicago and NBC Radio, Transcript: The Unversity of Chicago Round Table—The Facts about the Hydrogen Bomb. Part 2 of "How We Can Make Peace" (February 26, 1950)

Welsome, Eileen, *The Plutonium Files: America's Secret Medical Experiments in the Cold War*. New York, 1999 (Delta)

Wrong, Michela, *In the Footsteps of Mr. Kurtz: Living on the Brink of Disaster in the Kongo*. London, 2000 (Fourth Estate)

Zippe, Gernot, *Rasende Ofenrohre in stürmischen Zeiten: Ein Erfinderschicksal aus der Geschichte der Uranisotopentrennung im heißen und im kalten Krieg des 20. Jahrhunderts*. Vienna, 2008 (self-published)

Index

Davy Crocket rocket launcher test, 55, 80i
Ely, 104
Harry tests, 61–65
Sugar explosion, 68
underground tests in, 68, 71
New York Times, 113–114, 115, 117, 119–121, 123
New Zealand, 38
9/11 attacks, 44, 122
Nixon, Richard, 65, 215
Noatak, Alaska, 69, 93
North Carolina, 200–202
North Korea, 7, 28, 213, 216
Norway, 8, 9, 38
Novaya Zemlya archipelago, 35–40, 42
nuclear agreements, 34, 44, 107, 217
nuclear batteries, 127, 157, 191
nuclear earthmoving. *See* seismological and earthmoving tests
nuclear electricity, 159–160
Nuclear Emergency Search Team (NEST), 130
nuclear energy
    energy companies and, 160, 161, 163
    global demand for, 185, 217
    power plants and reactors for, 158–162
Nuclear Energy Commission (NEC), Brazil, 187–188
nuclear fission
    destructive power of, 24, 87, 114
    discovery of, 8
    energy released during, 10
    past optimistic views of, 5
nuclear landmines, 53, 104–105
nuclear medicine
    dangers of, 191
    Goiana disaster, 187–191
    human radiation experiments, 147i, 178–180, 193–196
    moral questions in, 193–194

    radioactivity and success in, 192–193
Nuclear Non-Proliferation Treaty, 7, 186
nuclear planes and jets, 134–137, 144i
nuclear power plants and reactors
    Chernobyl catastrophe, 4–5, 42, 50, 162–163
    civilian *vs.* military, 159
    in Congo, 153–157
    costs of developing, 160–162
    Fukushima disaster, 5, 42, 156–157, 163, 217
    in Germany, 164
    reactor's cores, 161
    TRIGA Mark I and II reactors, 154, 156
nuclear satellites, 127–134
nuclear warheads
    Chrome Dome missions, 199–204
    Mars Bluff tests, 187–199
nuclear weapons technology
    anti-nuclear protests against, 71–72, 109–110
    design of, 14–15, 150i
    destructive power and danger of, 12–13, 64–65, 88–89, 107, 141i, 158
    global spread of, 3–5, 7–9, 32–33, 49, 216–217
    miniaturization of, 56–57
    moral questions on, 158
    in Nazi Germany, 7–14
    in United Kingdom, 168–169
    See also hydrogen bombs
Nuclebrás, 190
Nucleon, 134, 143i
nylon, 90

O

Oga Bay, 37
Ogotoruk Creek, 93, 97–98, 99, 101
Oko nuclear early-warning system, 123

Acknowledgments

For their advice and help, I would like to thank Martje Herzog, Marco Wehr, Hans-Jürgen Bleif, Ulrich Zanke, Elisabeth Kraft, and Jens Zinglersen.

Picture Credits